Reproductive Biology of Plants

Reproductive Biology of Plants

Editor

Shalini Dwivedi

scitus
academics

Reproductive Biology of Plants

Edited by **Shalini Dwivedi**

ISBN: 978-1-68117-028-2

Library of Congress Control Number: 2015931389

SCITUS Academics LLC,
www.scitusacademics.com
Box No. 4766, 616 Corporate Way,
Suite 2, Valley Cottage,
NY 10989

Notice

Printed in United States of America on Acid Free Paper ∞

Contents

	Preface	vii
Chapter 1	**Structure, Function, and Floristic Relationships of Plant Communities in Stressful Habitats Marginal to the Brazilian Atlantic Rainforest** Fabio R. Scarano	1
Chapter 2	**Pollination and Reproductive Biology of Twelve Species of Neotropical Malpighiaceae: Stigma Morphology and its Implications for the Breeding System** Maria Rosângela Sigrist and Marlies Sazima	25
Chapter 3	**Effects of Pollination Limitation and Seed Predation on Female Reproductive Success of a Deceptive Orchid** Ryan P. Walsh, Paige M. Arnold and Helen J. Michaels	49
Chapter 4	**Habitat Specificity of A Threatened and Endemic, Cliff-Dwelling Halophyte** Ana D. Caperta, M. Dalila Espírito-Santo, Vasco Silva, Ana Ferreira, Ana P. Paes, Ana S. Róis, José C. Costa and Pedro Arsénio	79
Chapter 5	**Floral Specialization and Angiosperm Diversity: Phenotypic Divergence, Fitness Trade-Offs and Realized Pollination Accuracy** W. Scott Armbruster	107
Chapter 6	**A Functional Role for the Colleters of Coffee Flowers** Juliana Lischka Sampaio Mayer, Sandra Maria Carmello-Guerreiro, and Paulo Mazzafera	169
Chapter 7	**Structural Organization of the Gynoecium and Pollen Tube Path in Himalayan Sea Buckthorn, Hippophae Rhamnoides (Elaeagnaceae)** Yash Mangla, Rajesh Tandon, Shailendra Goel and S. N. Raina	193

Chapter 8 **Genetic Diversity in Napier Grass (Pennisetum Purpureum) Cultivars: Implications for Breeding and Conservation 215**

Bramwel W. Wanjala, Meshack Obonyo, Francis N. Wachira, Alice Muchugi, Margaret Mulaa, Jagger Harvey, Robert A. Skilton, Janice Proud, and Jean Hanson

Chapter 9 **Mineral Nutrient Remobilization During Corolla Senescence in Ethylene Sensitive and -Insensitive Flowers 237**

Michelle L. Jones

Chapter 10 **Characterization of TM8, a MADS-box Gene Expressed in Tomato Flowers 263**

Margherita Daminato, Simona Masiero, Francesca Resentini, Alessandro Lovisetto, and Giorgio Casadoro

Citations..295

Index..299

Preface

Petroleum Radiation Processing fills an information gap, providing systematic descriptions of the fundamentals of radiation-induced cracking reactions in hydrocarbons. It analyzes the basic experiments that have brought about the rapid development of radiation technology for petroleum radiation processing during the last decades. The book emphasizes high-viscous oil feedstocks that are difficult to process by conventional methods—such as heavy and high-paraffinic crude oil, fuel oil, and bitumen. It helps readers understand the mechanisms and kinetics of low-temperature radiation cracking. The book addresses the application of promising radiation methods for solving critical environmental issues, such as oil desulfurization and regeneration of used lubricants and other used oil products. Examining experimental data as well as theoretical and technical approaches, it summarizes research progress in the field of petroleum radiation processing, providing a useful reference on the theory and technology of hydrocarbon radiation processing for scientists, researchers, and students.

Editor

Chapter 1

Structure, Function, and Floristic Relationships of Plant Communities in Stressful Habitats Marginal to the Brazilian Atlantic Rainforest

Fabio R. Scarano*[1]

[1]Universidade Federal do Rio de Janeiro, CCS, IB, Departamento de Ecologia, Caixa postal 68020, cep 21941-970, Rio de Janeiro, RJ, Brazil

ABSTRACT

The Brazilian Atlantic rainforest consists of a typical tropical rainforest on mountain slopes, and stands out as a biodiversity hotspot for its high species richness and high level of species endemism. This forest is bordered by plant communities with lower species diversity, due mostly to more extreme environmental conditions than those found in the mesic rainforest. Between the mountain slopes and the sea, the coastal plains have swamp forests, dry semi-deciduous forests and open thicket vegetation on marine sand deposits. At the other extreme, on top of the mountains (>2000 m a.s.l.), the rainforest is substituted by high altitude fields and open thicket vegetation on rocky outcrops. Thus, the plant communities that are marginal to the rainforest are subjected either to flooding, drought, oceanicity or cold winter temperatures. It was found that positive interactions among plants play an important role in the structuring and functioning of a swamp forest, a coastal sandy vegetation and a cold, high altitude vegetation in the state of Rio de Janeiro. Moreover, only a few species seem to adopt this positive role and, therefore, the functioning of these entire systems may rely on them. Curiously, these nurse plants are often epiphytes in the rainforest, and at the study sites are typically terrestrial. Many exhibit crassulacean acid metabolism. Conservation initiatives must treat the Atlantic coastal vegetation as a complex rather than a rainforest alone.

INTRODUCTION

The Brazilian Atlantic rainforest once covered most of eastern Brazil. It has now been reduced to only 7·5 % of its original area. Since it stands out as one of the main centres of biodiversity in the world, with high species diversity and high levels of endemism, it is now considered from a global perspective as one of the 25 biodiversity hotspots for conservation priorities (Fonseca, 1985; WCMC, 1992; Myers *et al.*, 2000). The worldwide notoriety of the rainforest at the core of this complex hides the fact that it is only one of several types of plant communities within a much greater vegetation complex. For instance, a recent review of the Atlantic forests defines them as vegetation composed of two forest types: a coastal rainforest and a semi-deciduous forest type (Morellato and Haddad, 2000).Oliveira-Filho and Fontes (2000) proposed that the definition of Atlantic forests should be comprehensive, encompassing rainforests, semi-deciduous forests and also the southern subtropical *Araucaria* forests and the northeastern enclaves of *brejo* forests. My argument is that from evolutionary and conservation viewpoints, the Atlantic coastal vegetation of Brazil, particularly in the case of the state of Rio de Janeiro, should be treated as a mosaic comprising all forest types and also the neighbouring open vegetation. As long ago as 1979, Rizzini noted that the Atlantic coastal vegetation should be seen as a complex formed by several plant communities including those at the periphery of the forests.

Plant communities at the periphery of the Atlantic rainforest are subjected to more extreme environmental conditions than plants in the mesic rainforest (Scarano *et al.*, 2001). These communities experience a wide array of adverse environmental conditions, such as high and low (including freezing) temperatures, flooding, drought, constant wind, high salinity and lack of nutrients. Thus, diversity, productivity and structural complexity are lower in these communities. Table 1 lists the main characteristics of the plant communities found within the Atlantic coastal vegetation. Figure1 illustrates the geographic location of these communities in the state of Rio de Janeiro.

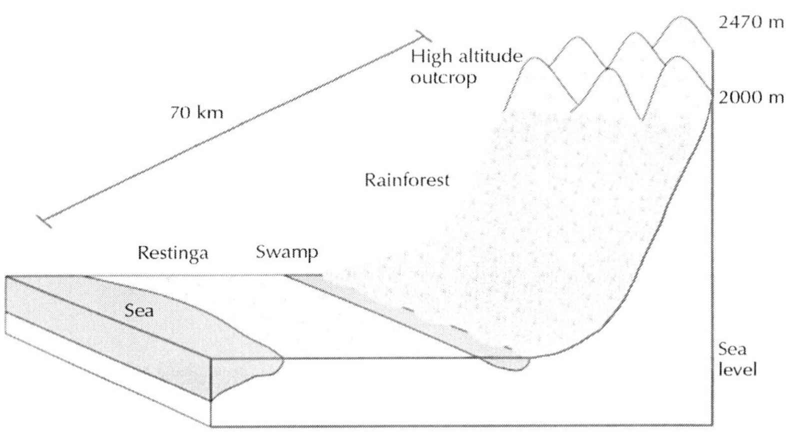

Figure 1: Schematic representation of the Atlantic forest complex in the state of Rio de Janeiro. Note that the distance between the lowland and the high altitude vegetation is only approx. 70 km as the crow flies.

Table 1: Some of the main plant communities of the Atlantic forest complex in the state of Rio de Janeiro, south-eastern Brazil

Habitat	Location	Limiting factors	References
High altitude	Includes *Araucaria* forest, marshes and rocky outcrops; >2000 m a.s.l.	Frost, drought, shallow or absent soil, high light irradiation	Safford (1999a, b)
Rainforest	Sea level to approx. 1500 m a.s.l.	Deep shade	Rizzini (1979)
Rocky outcrops	Coast and inland; elevations from 0-1400 m a.s.l.	Drought, shallow or absent soil, salinity (coast)	Meirelles *et aL* (1999)
Swamp forest	Coastal lowlands	Flooding	Scarano *et al.* (1997)
Dry forests	Coastal lowlands (mainly at Buzios-Cabo Frio region, north of Rio de Janeiro)	Drought, salinity, low in nutrients	Araujo (1997)
Restingas	Along the coast, includes marshes, dry and swamp forests and open clumped vegetation	Drought, salinity, low in nutrients	Lacerda *et al.* (1993)

This paper synthesizes a decade of research on the plant communities at the periphery of the Atlantic forest in the state of Rio de Janeiro. It covers aspects related to plant community structure, community function and phytogeography. It focuses on three vegetation types: the open scrub vegetation of the sandy coastal plains, the swamp forests and the vegetation growing on rocky outcrops at high altitude. Despite the extreme environmental differences between these habitats, I propose that community structure and function among these habitats is strikingly similar. On the basis of evidence presented in this paper and also the floristic and evolutionary links between marginal habitats and the Atlantic rainforest (Rizzini, 1979; Araujo, 2000; Lima, 2000), I also propose that conservation initiatives must treat the Atlantic forest as a vegetation complex rather than as a single entity, i.e. the forest formations.

STUDY SITES

This paper reviews studies conducted in three vegetation types: (*a*) swamp forests growing on periodically or permanently flooded terrain at the foot of the coastal mountain; (*b*) a mosaic of plant communities collectively known as the *restinga*, which occupies the sandy plains formed by marine deposits in the late Quaternary that stand between the sea and the Atlantic forest mountain chain; and (*c*) high altitude vegetation, on the mountain highlands above the treeline and ranging from 2000 to 2470 m a.s.l., consisting of fields covered by grasslands, marshes and vegetation on rocky outcrops. This review compares data collected for specific plant communities within each of these three vegetation types, in the state of Rio de Janeiro. The information in Table 2 briefly describes the sites where most of the research for this review was conducted.

Table 2: Characteristics of the Brazilian Atlantic Rainforest and associated habitats referred to in the text

Habitat	Location	Climate	Nurse plant species	References
High altitude rocky outcrops	Itatiaia National Park (20°25′S, 44°50′W; 2400 m a.s.l.)	Markedly seasonal. Annual rainfall of 2273 mm, concentrated in the summer (November to February). Winter (May to August): cold, dry, and less cloudy than the summer. Summer daytime maximum temperature of 23 °C and winter night-time minimum of –10 °C.	*Fernseea itatiaiae**, *Pleurostima gounelleana**, *Campylopus pilifer**, *Polytrichum commune**	Scarano *et al.* (2001); Ribeiro (2002)

Swamp forest	Poço das Antas Biological Reserve (22°30'S, 42°15'W; sea level)	2260 mm rainfall well distributed throughout the year, with a discrete dry season from May to August. Annual temperatures: mean = 25 °C, max = 38 °C, min = 14 °C.	*Nidularium procerum* †	Scarano *et al.* (1997); Rôçaset *al.* (2001)
Open restingas	Environmental Protection Area of Massambaba (22°56'S, 42°13'W; sea level)	Markedly seasonal. Annual rainfall of 800 mm concentrated in the summer (November to February). Annual temperatures: mean = 25 °C, max = 36 °C, min = 12 °C.	*Allagoptera arenaria**, *Neoregelia cruenta*†, *Clusia fluminensis*†	Zaluar and Scarano (2000); Scarano *et al.* (2001)
	Restinga de Jurubatiba National Park (22°23'S, 41°45'W; sea level)	Markedly seasonal. Annual rainfall of 1164 mm concentrated in the summer (November to February). Annual temperatures: mean = 23 °C, max = 30 °C, min = 20 °C.	*Allagoptera arenaria**, *Aechmea nudicaulis*†, *Clusia hilariana*†	Zaluar and Scarano (2000); Liebiget *al.* (2001)

* Both pioneer and nurse plants.

† Nurse plants only.

There is still little detailed and uncontroversial information about the palaeo-ecological, geological and recent environmental history of the study sites and of the Atlantic complex in Rio de Janeiro as a whole. The rainforests in the state of Rio de Janeiro date from the Tertiary and cover areas with annual rainfall as low as 1173 mm; rainfall is, however, well-distributed (Oliveira-Filho and Fontes, 2000). In the north of Rio de Janeiro state the mountains are further from the coast than they are in the south (Oliveira-Filho and Fontes, 2000), and the north therefore has more extensive and wider lowland plains (see Fig. 1). The swamps stand on a substrate that is composed mainly of Quaternary alluviums, and are punctuated by rounded hills ranging from 30 to 205 m. Vegetation on the hills is typically a sub-montane rainforest. Forest cover of swamps in Rio de Janeiro has been reduced to a minimum since the mid-20th century due to river dam construction and wood extraction. The data discussed here were obtained at the Poço das Antas Biological Reserve, where only 50 % of the total area consists of well-preserved remnants (Lima, 2000).

Most sandy coastal plains in the state of Rio de Janeiro consist of two beach ridges dating from the Holocene [5000 to 3000 years before the present (BP)]. The oldest sandy plains established in the state date from the Pleistocene (120 000 years BP; see Martin *et al.*, 1993). Araujo (2000) classified the restingas of the state of Rio de Janeiro into ten zones according to their flora, which is related to local variation in climate and geological history. The data reviewed here were obtained in two of these zones, both located on the northern coast of the state. The climate here is drier due to the cold oceanic upwelling of the Cabo Frio region (Araujo, 1997). One of the sites examined is within the Cabo Frio region (Environmental Protection Area of Massambaba), and the other (Restinga de Jurubatiba National Park) is approx. 100 km further north (see also Table 2).

The Massambaba sandy plains stretch between the ocean and the pleistocenic lagoon called Araruama, which is Brazil's second largest salt lake. They are composed of two parallel beach ridges of different ages, separated by a low area, where holocenic lagoons showing different sedimentation stages are situated (Muehe, 1994). On its east side, where most of the data reviewed here were collected, the pleistocenic, innermost beach ridge (Martin and Suguio, 1989) was covered by a dune system approx. 2000 years BP (FEEMA, 1988). Here, the distance between the ocean and the Araruama lagoon is approx. 1·5 km (Scarano *et al.*, 2001).

Unlike most sandy coastal plains in Rio de Janeiro, the Jurubatiba plains consist of a series of parallel pleistocenic beach ridges (120 000 years BP) formed by the silting of marine sediments, which formed several lagoons due to the obstruction of ancient river deltas (Henriques *et al.*, 1986). The series of beach ridges reach a maximum width of

10 km. Along the Brazilian coast, three successive transgression movements of the sea took place during the Holocene (approx. 5100 years BP, 3900–3600 years BP and 2700–2500 years BP) followed by regressions (Martin *et al.*, 1989). However, the length of transgression was progressively smaller here and apparently did not destroy the Pleistocene ridges completely.

The high altitude vegetation in Rio de Janeiro state has recently been reviewed with regard to climatic, physical and biotic characteristics (Safford, 1999*a*, *b*). The data described here refer mostly to the rocky outcrops within the high altitude fields of the Itatiaia Plateau, at the Itatiaia National Park (see also Table 2). Whether or not glaciers occurred in this region during glaciation peaks is still the subject of debate (Ebert, 1960;Segadas-Vianna, 1965). However, a temperature regime similar to the current one in the plateau (Table 2) may have occurred at altitudes 1000 m lower, while peri-glacial conditions reigned in the plateau itself (Behling, 1998). Rocky outcrops dominate the landscape of the plateau and the predominant rock is nepheline-sienite, an alkaline pluton not very common in Brazil. It has a high dissolution rate, and channels and holes form frequently (Leinz and Amaral, 1989). However, narrow fissures, which are common in granite and favour the fixation of plant species, are rare. The occurrence of soil on the rocks is also rare, patchy and shallow.

PHYTOGEOGRAPHY AND FLORISTIC RELATIONSHIPS

Restingas, swamps and high altitude vegetation have distinct floristic links with the neighbouring Atlantic rainforest. While most of the flora of the restinga (Rizzini, 1979) and the swamp forest (Lima, 2000) originated from the Atlantic rainforest, the flora of the high altitude rocky outcrop vegetation bears less resemblance to it (Safford, 1999*a*, *b*; Ribeiro, 2002).Araujo (2000) showed that approx. 60 % of the species listed for the restinga of coastal Rio de Janeiro occur in the Atlantic rainforest; the remainder originated in other vegetation of Rio de Janeiro (coastal inselbergs and semi-deciduous dry forests) and elsewhere in Brazil (cerrado, caatinga and the Amazon). Oliveira-Filho and Ratter (1995) showed that many rainforest species have expanded their distribution into markedly seasonal habitats via riverside forests; this may also apply to the restingas. Very few plant species seem to be endemic to this vegetation. The same is true for birds and mammals: 98 % of restinga birds are found in the rainforest (Reis, 1998) and restinga mammals are a mere sub-group of the same fauna from the rainforest (Cerqueira *et al.*, 1990).

To date, the swamps have not been subjected to a thorough phytogeographic study (but see Lima, 2000), but some of the species lists available indicate that the flora is mixed, being mostly composed of: (*a*) a few typically flood-tolerant trees broadly distributed in other parts of Brazil, such as *Calophyllum brasiliense* Cambess and *Symphonia globulifera* L.f. (both Clusiaceae); (*b*) a few flood-tolerant trees restricted to the southeast of the country, e.g. *Tabebuia cassinoides* (Lam.) D.C. (Bignoniaceae); and (*c*) many less abundant tree species that also have their origins in the Atlantic rainforest (Scarano *et al.*, 1998).

The almost total absence of endemic plant species in the lowlands suggests that, despite the differences in terms of selective pressures between these peripheral habitats and the rainforest, there has been insufficient time for plant speciation. Perhaps even more surprising is the fact that species richness in the lowlands is quite high. This implies that a large number of rainforest plant species were able to adjust to extreme conditions in order to colonize the geologically younger lowlands. Since new species could not have evolved during this time, it is likely that the successful migrating plant species from the Atlantic rainforest had a broad ecological plasticity. Recent studies have indicated wide inter- and intrapopulation phenotypic variation for some of these species, both locally and across habitats, with regard to architecture, leaf and wood morphology and anatomy, physiology and growth patterns (Freitas *et al.*, 1998; Herzog *et al.*, 1999; Callado *et al.*, 2001*a*, *b*; Cirne and Scarano, 2001; Rôças *et al.*, 2001; Scarano *et al.*, 2001), which is evidence of such ecological plasticity.

Historically, the high altitude vegetation has offered several opportunities for the fixation and differentiation of forest, cerrado (the neotropical savanna) and temperate species (Brade, 1956). For instance, during the Pleistocene glaciations, the high altitude areas were probably nearer to distinct kinds of open vegetation, while forests were retracting (Ribeiro, 2002). Climatic oscillations could also explain its floristic disjunctions regarding the Andes (Brade, 1956; Safford, 1999*a*), and the grassland vegetation of southern Brazil (Brade, 1956). On the other hand, the current border with the Atlantic forest is a possible port of entry for tropical species in a temperate environment (Ribeiro, 2002). However, the high altitude flora in the state of Rio de Janeiro currently differs floristically from the rainforest at lower altitudes in physiognomy and floristics, mostly due to soil scarcity and patchiness (see Study Sites) and climate (Safford, 1999*a*, *b*; Ribeiro, 2002). Temperatures below 0 °C are common on winter nights on the Itatiaia Plateau, where an average of 56 frosts occur per year (Segadas-Vianna, 1965; Segadas-Vianna and Dau, 1965). These low temperatures provide a strong barrier for establishment of most taxa from the rainforest, which are predominantly tropical (Gentry, 1982). Temperatures between 6 and 10 °C are sufficient to cause chilling injury, whereas those below 0 °C cause freezing injury in most tropical plants (Crawford, 1989). Thus, low temperatures have led to selection for a high proportion of local endemics and species endemic to mountain peaks (Safford, 1999*a*; Ribeiro, 2002). Despite the high level of endemism,Scarano *et al.* (2001) have shown that some of these species are apparently quite plastic in terms of carbon, water and nitrogen use. For instance, *Fernseea itatiaiae* (Wawra) Baker, a local endemic to the high altitude rocky outcrops, showed the highest variation of $\delta^{15}N$ values among the approx. 30 species investigated from restingas, dry forests and high altitudes. Values ranged from –2·52 to +1·81 ‰ (4·3 ‰ variation). This might be related to variation in root system functioning in response to soil depth, before reaching bare rock.

COMMUNITY STRUCTURE AND FUNCTION

Other than the obvious structural difference between swamp forests and open, scrub vegetation of restingas and high altitude rocky outcrops, these three marginal vegetation types are strikingly similar in structure and function. They are species-rich but are dominated by one (swamp) or a few (open vegetation types) species. Monodominance and oligarchy are common features in habitats subjected to extreme environmental conditions. Additionally, the functioning of these vegetation types appears to rely on a few species that provide appropriate environmental conditions for germination and growth of other species, the so-called 'nurse plants' (Franco and Nobel, 1989; Young *et al.*, 1995). An overview of studies conducted on the swamp forests (Scarano *et al.*, 1997, 1998, 1999; Freitas*et al.*, 1998) suggests that it can be divided into three functional groups related to characteristic plant interactions: (*a*) the monodominant, emergent, deciduous tree, *Tabebuia cassinoides* DC (Bignoniaceae), which does not provide year-round shade to understorey plants; (*b*) 59 tree species represented by a small number of individuals (approx. 30 % of the total, on 0·5 ha) that are mostly evergreen, and provide shade to the understorey throughout the year; and (*c*) the monodominant understorey herb, the light-sensitive tank-bromeliad *Nidularium procerum* Lindm., which provides safe germination sites (wide leaf tanks that trap litter and seeds) for plants that would not germinate on the permanently flooded soil below them. Although there is a lack of long-term monitoring of water-level fluctuations in the swamp, it is possible that in drier years massive seed germination might take place in the soil since the water layer is only superficial. In spite of this, the tank-bromeliad plays a vital role for the locally rare trees by providing year-round germination sites, while the rare trees provide year-round shade for the light-sensitive bromeliads. The monodominant tree *T. cassinoides* does not seem to have any positive interaction with the locally rare trees or with the bromeliads. Although seeds of *T. cassinoides* may germinate in the 'suspended soil' provided by the bromeliads, it does not rely on this exclusively since it is able to reproduce vegetatively. Scarano *et al.* (1998) proposed that this functional organization is the reason why degraded swamp forests often turn into exclusive stands of *T. cassinoides*: removal of the shade-providing locally rare trees results in the disappearance of bromeliads and *vice versa,* whereas the *T. cassinoides* population remains unaffected. It remains to be seen how these bromeliads are capable of withstanding long-term flooding. The very small rhizomes of these plants may rule out the possibility of tolerance mechanisms. Freitas *et al.* (1998) proposed that these plants behave as 'epiphytes on mud'. Shoot autonomy for foraging and the formation of tank roots, i.e. geotropically negative roots that forage into the leaf rosette above them (see Pittendrigh, 1948; Medina, 1974), probably constitute an efficient escape mechanism on the flooded ground, as they do in epiphytic habitats for these bromeliads.

Positive plant–plant interactions, such as those described for the swamp forest, are also seen in the open vegetation types. Zaluar and

Scarano (2000) reviewed colonization and vegetation dynamics in the open scrub vegetation of the restingas. There has been much debate as to why this vegetation is open and there is a lack of palaeo-ecological evidence in this respect. However, it is known that human occupation of the sandy plains of Rio de Janeiro dates from 8000 years BP (Kneip and Pallestrini, 1984), but more dramatic changes in the landscape have probably taken place since the country was discovered by the Portuguese in the year 1500. Today, some restingas still suffer man-made impact through fire (Cirne and Scarano, 2001) or cattle, but even apparently pristine areas display an open physiognomy.

In the restingas, formation of vegetation islands and succession are triggered by the colonization of pioneer herbs and/or woody plants, which may favour the entry of other species into this community. The ability to colonize bare sand (which may reach temperatures as high as 70 °C during mid-summer at the peak of radiation) via seeds is restricted to a few species such as the geophyte palm *Allagoptera arenaria* (Gomes) Kuntze. Seeds of drought-resistant plants such as cacti and bromeliads require shade provided by shrubs to germinate. Small vegetation islands formed by the palm, cacti [e.g. *Pilosocereus arrabidae* (Lem.) Byles & G.D. Rowley and *Cereus fernambucensis* Lem.] and bromeliads [e.g. *Aechmea nudicaulis* (L.) Griseb and *Neoregelia cruenta* (Graham) L.B. Sm.] favour the entry of woody species. Among the woody species, those belonging to the genus *Clusia* (e.g. *C. hilariana* Schlecht. and *C. fluminensis* Planch & Triana) seem to be the main nurse plants under which many plant species become established (Zaluar and Scarano, 2000). However, other woody species may form vegetation islands where few or no other species become established [e.g. *Protium icicariba* (DC.) Marchand]. In some cases, such as in the Restinga de Jurubatiba National Park, *P. icicariba* and *C. hilariana* share community dominance. Thus, neighbouring vegetation islands of a given restinga may have totally different species richness, diversity and structure depending on the dominant woody species of the vegetation island. They may also be in different trajectories along the successional process. Small islands can be either newly formed or simply an old fragment of a larger island, which retracted due to the death of its nurse tree. Large islands can also result from fusion of several smaller islands. In the future, analyses of time series of aerial photographs, along with species addition and removal experiments, will allow a better understanding of the dynamics and succession of vegetation islands in the restingas. Unlike the restingas, the openness of the vegetation on rocky outcrops of Itatiaia has not been attributed to long-term human intervention, but rather to the inherent difficulties related to plant fixation on rocks or shallow soil, as discussed previously. However, the study area is not totally free from current human intervention, since predatory tourism (on a more regular basis) and man-made fires (less frequent) occur. Similarly to the open restinga vegetation, it has been determined that plants which form mats on the rocky surfaces of the high altitude outcrops are nurse plants that provide the substrate upon which other plants establish (Ribeiro, 2002). Common pioneer mat species at altitude in Itatiaia are *Fernseea itatiaiae* (Bromeliaceae), *Pleurostima gounelleana* (Beauv) N.L.de Menezes (Velloziaceae), *Campylopus pilifer* Brid. and *Polytrichum commune* Hedw. (both Musci). Geophytes such as *Stevia camporum* Baker (Asteraceae) and*Alstroemeria foliosa* Mart. (Alstroemeriaceae) grow on mat species and not on the bare rock. The former is more frequent on islands of mosses, whereas the latter is more abundant on islands of *Pleurostima*. Although variations in relief on the rocky outcrops provide a myriad of microhabitats, habitat preference of these geophytes is dependent upon the mat species. Successional mechanisms on the high altitude rocky outcrops are therefore more decisively triggered by species–species relationships than by topography.

RAINFOREST CANOPY PLANTS AS TERRESTRIAL NURSE PLANTS IN THE MARGINAL HABITATS OF THE FOREST

Table 2 summarizes the species that play a role as nurse plants in the plant communities studied. Typically, most of the nurse plants are epiphytes in the Atlantic rainforest, whereas in the marginal habitats they are terrestrial plants playing a role as nurse plants. This is true of the bromeliads in the swamp forests (*Nidularium procerum*), in the restingas (*Neoregelia cruenta* and *Aechmea nudicaulis*) and in the high altitude rocky outcrops (*Fernseea itatiaiae*). Similarly, the *Clusia* plants in the restinga are hemiepiphytic stranglers in the rainforest.

These species, with the apparent exception of *Fernseea* (Scarano *et al.*, 2001), are CAM (crassulacean acid metabolism) plants (Franco *et al.*, 1996; Reinert *et al.*, 1997; Lüttge, 1999; Scarano *et al.*, 1999).

CAM is characteristically a stress-resistance mechanism that maximizes water-use efficiency. Although, it is not surprising that stress-resistant plants may nurse less resistant ones, the fact that these terrestrial clusias and bromeliads have CAM metabolism and originate in the montane Atlantic rainforest is of great interest. Their migration to the geologically younger lowlands of the restingas and swamp forests was seemingly followed by the expression of a new habit (from epiphytic in the rainforest to terrestrial plants in the lowlands) and a new ecological function (nursing seeds and seedlings of other plant species). Thus, the rainforest strangler*Clusia* is the most dominant nurse shrub of the open restingas, and the epiphytic bromeliads of the rainforest provide safe germination sites as terrestrial plants in inhospitable substrates such as the hot sandy soil of the restingas and the flooded soil of the swamps. Although CAM is not a common feature of plants on the high altitude rocky outcrops (see Scarano*et al.*, 2001), the main nurse plants (bromeliads, Velloziaceae and musci) are either epiphytic in forests or linked to phylogenetic groups where epiphytism is common. Porembski *et al.* (1998) proposed that the high plant species diversity found in inselbergs within the Atlantic forest domain is partially due to exchange of species between the rocky habitats and the canopy of neighbouring forests.

Medina (1974) discussed the question of whether forest epiphytes originated from terrestrial parents in open, sun-exposed vegetation or if epiphytes migrated from the forest floor to the canopy, and then subsequently colonized open vegetation. The latter possibility seems more plausible for the geologically younger lowlands and sandy plains, whereas the former possibility may have occurred at high altitude. In the case of the Atlantic forest complex, then, both trajectories seem likely.

Forest canopy plants have apparently played a major role in increasing species richness in these marginal habitats. In the high altitude outcrops, the nurse plants are also pioneers, capable of germinating and growing on almost bare rock; this does not seem to be the case in the lowlands: clusias and bromeliads are unable to germinate on bare sand or on flooded soil. The pioneer plant in the restingas is predominantly the geophyte palm *Allagoptera arenaria*. Establishment is therefore a major hurdle to be overcome by plants in such habitats, and only a few can do so without the aid of another species. Thus, stress-resistance through mechanisms such as CAM (Herzog *et al.*, 1999; Scarano *et al.*, 2001) or flood-tolerance (Scarano *et al.*, 1997) are not sufficient to ensure establishment in restingas and swamps.

CONSERVATION

The plant communities marginal to the Atlantic rainforest are structurally and functionally dependent on a small number of nurse plants. This small number of nurse plants, in an otherwise species-rich community, often results in a smaller degree of functional redundancy and in higher community fragility (see Shrader-Frechette and McCoy, 1993). The eventual loss of nurse plants in such communities may result in dramatic changes in species composition and diversity, as shown by Scarano *et al.* (1998) for the swamp forest. Experimental removal of nurse plants in similarly structured communities often results in major changes in species composition (e.g. Holzapfel and Mahall, 1999).

The functional ecology of plant communities is often not a part of conservation initiatives, which are mostly based on numerical diversity, levels of community endemism and rates of habitat destruction. In the Atlantic forest complex, an apparent dualism exists between the rainforest, which fits these criteria, and its marginal habitats, which are less diverse and show less endemism (although the high altitude rocky outcrops are a notable exception). For instance, the classification of biodiversity hotspots provided by Myers *et al.* (2000) does not include these marginal habitats. Unfortunately, the distribution of federal conservation units in Brazil is also more concentrated on 'hotspots', e.g. only 14 000 ha of restingas are protected as the single federal conservation unit for restinga in the country (Restinga de Jurubatiba National Park).

Smith *et al.* (2001) pointed out that bias might arise from the biodiversity hotspot classification, resulting in a disregard for important evolutionary and adaptive processes taking place in low diversity communities and/or ecotones. Thus, the eventual establishment of conservation goals or research priorities as guided by such a classification could be misleading. This could be the case for the Atlantic rainforest complex, if viewed solely as the rainforest and not inclusively as the rainforest plus its marginal habitats. This is also true with regards to animal conservation. Although fragmentation has often removed connections between the rainforest and its marginal habitats, many bird species from the rainforest are reported to use the restingas as a buffer zone and as a geographic extension of their distribution (Gonzaga *et al.*, 2000). In the case of mammals, the only two studies on large-scale movements in the landscape (Pires and Fernandez, 1999; Pires *et al.*, 2002) report movements of marsupials and rodents between isolated forest fragments in the Atlantic forest lowlands where the swamps are located.

Perhaps more importantly than endemism, the rainforest has a pool of highly environmentally plastic species which have the capacity to colonize habitats distinctly different from their own, such as restingas and swamps. This ecological plasticity expressed in plants at these marginal habitats may be of key relevance in a global change scenario and must not be treated as a lower conservation priority.

FINAL REMARKS

This overview has presented an evolutionary hypothesis about the origins, structure and functioning of some plant communities that are marginal to the Atlantic rainforest of Rio de Janeiro, and a viewpoint regarding conservation of this vegetation complex. To further advance this hypothesis and to provide more subsidies for conservation, some key aspects must be addressed in the future: (*a*) palaeoecological studies, particularly of the Quaternary (e.g. Behling, 1998; Ledru *et al.*, 1998), are required to provide further insights into floristic and ecological links between the distinct plant communities; (*b*) DNA phylogenetic studies (Vaasen *et al.*, 2002) are needed to provide a molecular basis for the understanding of processes such as migration, speciation, vicariance and phenotypic plasticity; (*c*) studies on population genetics (e.g. Cardoso *et al.*, 1998) coupled with reproductive biology (e.g. Wendt *et al.*, 2001, 2002) are essential to assess genetic diversity and mating systems of nurse plants; and finally (*d*) experimental transplantation of species between habitats (e.g. Freitas *et al.*, 1998) and addition and removal of species within habitats will allow a more accurate assessment of the functional role of nurse plants and also of species plasticity.

ACKNOWLEDGEMENTS

I thank: D. S. D. Araujo, C. F. Barros, C. H. Callado, C. G. Costa, A. C. Franco, H. C. Lima, U. Lüttge, E. A. Mattos, M. T. Nascimento and T. Wendt for their collaboration and inspiring discussions on this topic; over 20 research students for taking part in the studies reviewed here; D. Biesboer, G. Durigan, E. A. Mattos and T. Wendt for valuable comments on the manuscript; H. L. T. Zaluar and K. T. Ribeiro for information on palaeo-ecology and recent environmental history and for insights on the nurse plant syndrome; M. A. S. Alves for information on bird ecology; F. A. S. Fernandez for information on mammal ecology; T. Wendt for graphic work in Fig. 1; IBAMA and FEEMA for granting access to conservation units; the Brazilian Research Council (CNPq) and Education Council (CAPES) for research grants; PRONEX-FINEP, PROBRAL-CAPES/DAAD, Brazilian LTER-CNPq (site 5), CTPetro-FINEP, Fundação O Boticário, the Brazilian Branch of the Margaret Mee Amazon Trust and FAPERJ for sponsoring the studies reviewed here; and PRONEX-FINEP (0877) for sponsoring my participation in Professor R. M. M. Crawford's retirement symposium where this paper was presented. Special thanks are due to Professor Crawford for constant advice and inspiration.

REFERENCES

1. AraujoDSD.1997. Cabo Frio region, south-eastern Brazil. In: Davis SD, Heywood VH, Herrera-Macbryde O, Villa-Lobos J, Hamilton AC, eds. *Centres of plant diversity: a guide and strategy for their conservation 3. The Americas*. Washington: WWF/IUCN,373–375.
2. AraujoDSD.2000. *Análise florística e fitogeográfica das restingas do Estado do Rio de Janeiro*. PhD Thesis, Universidade Federal do Rio de Janeiro, Brazil.
3. BehlingH.1998. Late Quaternary vegetation and climatic changes in Brazil.Review of Paleobotany and Palynology99: 143–156.

4. BradeAC.1956. A flora do Parque Nacional do Itatiaia. Boletim do Parque Nacional do Itatiaia5.
5. CalladoCH, Silva Neto SJ, Scarano FR, Costa CG.2001*a*. Periodicity of growth rings in some flood-prone trees of the Atlantic Rain Forest in Rio de Janeiro, Brazil.Trees Structure and Function15: 492–497.
6. CalladoCH, Silva Neto SJ, Scarano FR, Barros CF, Costa CG.2001*b*. Anatomic features of growth rings in flood-prone trees of the Atlantic Rain Forest in Rio de Janeiro, Brazil. IAWA Journal22: 29–42.
7. CardosoMA, Provan J, Powell W, Ferreiras PCE, de Oliveira PE.1998. High genetic differentiation among remnant populations of the endangered *Caesalpinia echinata* Lam. (Leguminosae-Caesal pinoideae). Molecular Ecology 7: 601–608.
8. CerqueiraR, Fernandez FAS, Quintela MFS.1990. Mamíferos da restinga de Barra de Maricá, Rio de Janeiro. Papéis Avulsos de Zoologia (São Paulo)47: 141–157.
9. CirneP, Scarano FR.2001. Resprouting and growth dynamics after fire of the clonal shrub *Andira legalis* (Leguminosae) in a sandy coastal plain in southeastern Brazil. Journal of Ecology 89: 351–357.
10. CrawfordRMM.1989. *Studies in plant survival*. Oxford: Blackwell.
11. EbertH.1960. Novas observações sobre a glaciação pleistocênica na Serra do Itatiaia. Anais da Academia Brasileira de Ciências32: 51–73.
12. FEEMA–Fundação Estadual do Meio Ambiente.1988. *Importância da biota da região de Cabo Frio*. Rio de Janeiro.
13. FonsecaGAB.1985. The vanishing Brazilian Atlantic forest. *Biological Conservation*34: 17–43.
14. FrancoAC, Nobel PS.1989. Effect of nurse plants on the microhabitat and growth of cacti. Journal of Ecology77: 870–886.
15. FrancoAC, Haag-Kerwer A, Herzog B, Grams TEE, Ball E, Mattos EA, Scarano FR, Barreto SMB, Garcia MA, Mantovani A, Lüttge U.1996. The effect of light levels on daily patterns of chlorophyll fluorescence and organic acid accumulation in the tropical CAM tree *Clusia hilariana*. Trees Structure and Function10: 359–365.
16. FreitasCA, Scarano FR, Wendt T.1998. Habitat choice in two

facultative epiphytes of the genus *Nidularium* (Bromeliaceae). Selbyana 19: 236–239.

17. GentryAH.1982. Neotropical floristic diversity: phytogeographical connections between Central and South America, Pleistocene climatic fluctuations, or an accident of the Andean orogeny? Annals of the Missouri Botanical Garden69: 557–593.
18. GonzagaLP, Castiglioni GD, Reis HBR.2000. Avifauna das restingas do sudeste: estado do conhecimento e potencial para futuros estudos. In: Esteves FA, Lacerda LD, eds. *Ecologia de Restingas e Lagoas Costeiras*. Rio de Janeiro: NUPEM-UFRJ, 151–163.
19. HenriquesRPB, Araujo DSD, Hay JD.1986. Descrição e classificação dos tipos de vegetação da restinga de Carapebus, Rio de Janeiro. Revista Brasileira de Botânica9:173–189.
20. HerzogB, Lüttge U, Hübner C, Ball E, Bastos RN, Franco AC, Scarano FR.1999. Comparative study of the C_3/CAM intermediate species *Clusia parviflora* Saldanha et Engl. and the obligate CAM species *Clusia hilariana* Schlecht. growing sympatrically exposed and shaded in the Atlantic restinga of Brazil. Plant Biology 1: 453–459.
21. HolzapfelC, Mahall BE.1999. Bidirectional facilitation and interference between shrubs and annuals in the Mojave desert. Ecology80: 1747–1761.
22. KneipLM, Pallestrini L.1984. Restingas do Estado do Rio de Janeiro (Niterói e Cabo Frio): 8 mil anos de ocupação. In: Lacerda LD, Araujo DSD, Cerqueira R, Turcq B, eds. *Restingas: origem, estrutura, processos*. Niterói: CEUFF, 139–155.
23. LacerdaLD, Araujo DSD, Maciel NC.1993. Dry coastal ecosystems of the tropical Brazilian coast. In: van der Maarel E, ed. *Dry coastal ecosystems: Africa, America, Asia and Oceania*. Amsterdam: Elsevier, 477–493.
24. LedruMP, Salgado-Labouriau ML, Lorscheiter ML.1998. Vegetation dynamics in southern and central Brazil during the last 10 000 yr B.P. Review of Paleobotany and Palynology99: 131–142.
25. LeinzV, Amaral SE.1989. *Geologia Geral*. São Paulo: Companhia Editora Nacional.

26. LiebigM, Scarano FR, Mattos EA, Zaluar HLT, Lüttge U.2001. Ecophysiological and floristic implications of sex expression in the dioecious neotropical CAM tree *Clusia hilariana* Schltdl. Trees Structure and Function 15: 278–288.

27. LimaHC.2000. *Leguminosas arbóreas da Mata Atlântica – Uma análise da riqueza, padrões de distribuição geográfica e similaridades florísticas em remanescentes florestais do Estado do Rio de Janeiro.* PhD Thesis, Universidade Federal do Rio de Janeiro, Brazil.

28. LüttgeU.1999. One morphotype, three physiotypes: sympatric species of *Clusia*with obligate C_3 photosynthesis, obligate CAM and C_3-CAM intermediate behaviour.Plant Biology 1: 138–148.

29. MartinL, Suguio K.1989. Excursion route along the Brazilian coast between Santos (state of São Paulo) and Campos (northern state of Rio de Janeiro).*International Symposium on Global Changes in South America during the Quaternary: Past, Present and Future.* São Paulo: special publication no. 2.

30. MartinL, Suguio K, Flexor JM.1993. As flutuações de nível do mar durante o quaternário superior e a evolução geológica de 'deltas' brasileiros. Boletim do Instituto de Geografia da Universidade de São Paulo, Publicação Especial 15: 1–186.

31. MedinaE.1974. Dark CO_2 fixation, habitat preference and evolution within the Bromeliaceae. Evolution 28: 677–686.

32. MeirellesST, Pivello VR, Joly CA.1999. The vegetation of granite outcrops in Rio de Janeiro, Brazil, and the need for its protection. Environmental Conservation26: 10–20.

33. MorellatoLPC, Haddad CFB.2000. Introduction: the Brazilian atlantic forest.Biotropica32(4b): 786–792.

34. MueheD.1994. Lagoa de Araruama: geomorfologia e sedimentação. Cadernos de Geografia do IBGE10: 53–62.

35. MyersN, Mittermeier RA, Mittermeier CG, Fonseca GAB, Kent J.2000. Biodiversity hotspots for conservation priorities. Nature403: 853–858.

36. Oliveira-FilhoAT, Fontes MAL.2000. Patterns of floristic differentiation among atlantic forests in southeastern Brazil and the influence of climate. Biotropica32(4b):793–810.

37. Oliveira–FilhoAT, Ratter JA.1995. A study of the origin of central

Brazilian forests by the analysis of plant species distribution patterns. Edinburgh Journal of Botany52: 141–194.

38. PiresAS, Fernandez FAS.1999. Use of space by the marsupial *Micoureus demerarae* in small Atlantic forest fragments in southeastern Brazil. Journal ofTropical Ecology 15: 279–290.

39. PiresAS, Lira PK, Fernandez FAS, Schittini GM, Oliveira LC.2002. Frequency of movements of small mammals among Atlantic coastal forest fragments in Brazil.Biological Conservation (in press).

40. PittendrighCS.1948. The bromeliad-*Anopheles*-malaria complex in Trinidad. I. The bromeliad flora. Evolution2: 58–59.

41. PorembskiS, Martinelli G, Ohlemuller R, Barthlott W.1998. Diversity and ecology of saxicolous vegetation mats on inselbergs in brazilian Atlantic forest. Diversity and Distributions4: 107–119.

42. ReinertF, Roberts A, Wilson JM, Ribas L, Cardinot G, Griffiths H.1997. Gradation in nutrient composition and photosynthetic pathways across the restinga vegetation of Brazil. Botanica Acta110: 135–142.

43. ReisHBR.1998. *Análise da composição e da distribuição geográfica da avifauna das restingas do Estado do Rio de Janeiro*. MSc Thesis, Universidade Federal do Rio de Janeiro, Brazil.

44. RibeiroKT.2002. *Estrutura, dinâmica e biogeografia das ilhas de vegetação rupícola no Planalto do Itatiaia, RJ*. PhD Thesis, Universidade Federal do Rio de Janeiro, Rio de Janeiro, Brazil.

45. RizziniCT.1979. *Tratado de fitogeografia do Brasil. Vol. 2*. São Paulo: Edusp.

46. RôçasG, Scarano FR, Barros CF.2001. Variation in leaf anatomy of *Alchornea triplinervia* (Spreng.) Müll. Arg. (Euphorbiaceae) under different light and soil-water regimes. Botanical Journal of the Linnean Society 136: 231–238.

47. SaffordHD.1999*a*. Brazilian Páramos I. An introduction to the physical environment and vegetation of the *campos de altitude*. Journal of Biogeography26:693–712.

48. SaffordHD.1999*b*. Brazilian Páramos II. Macro- and mesoclimate of the *campos de altitude* and affinities with high mountain

climates of the tropical Andes and Costa Rica. Journal of Biogeography 26: 713–737.

49. ScaranoFR, Rios RI, Esteves FA.1998. Tree species richness, diversity and flooding regime: case studies of recuperation after anthropic impact in Brazilian flood-prone forests. International Journal of Ecology and Environmental Sciences24:223–235.

50. ScaranoFR, Ribeiro KT, Moraes LFD, Lima HC.1997. Plant establishment on flooded and unflooded patches of a freshwater swamp forest of southeastern Brazil.Journal ofTropical Ecology14: 793–803.

51. ScaranoFR, Duarte HM, Ribeiro KT, Rodrigues PJFP, Barcellos EMB, Franco AC, Brulfert J, Deléens E, Lüttge U.2001. Four sites with contrasting environmental stress in southeastern Brazil: relations of species, life form diversity, and geographical distribution to ecophysiological parameters. Botanical Journal of the Linnean Society136: 345–364.

52. ScaranoFR, Mattos EA, Franco AC, Herzog B, Ball E, Grams TEE, Mantovani A, Barreto S, Haag-Kerwer A, Lüttge U.1999. Habitat segregation of C_3 and CAM*Nidularium* (Bromeliaceae) in response to different light regimes in the understory of a swamp forest in southeastern Brazil. Flora 194: 281–288.

53. Segadas-ViannaF.1965. Ecology of the Itatiaia range, southeastern Brazil. I – altitudinal zonation of the vegetation. Arquivos do Museu Nacional 53: 7–30.

54. Segadas-ViannaF, Dau L.1965. Ecology of the Itatiaia range, southeastern Brazil. II – climates. Arquivos do Museu Nacional 53: 31–53.

55. Shrader-FrechetteKS, McCoy ED.1993. *Method in ecology*. Cambridge: Cambridge University Press.

56. SmithTB, Kark S, Schneider CJ, Wayne RK, Moritz C.2001. Biodiversity hotspots and beyond: the need for preserving environmental transitions. Trends in Ecology and Evolution16: 431.

57. VaasenA, Begerow D, Lüttge U, Hampp R.2002. The genus *Clusia* L.: molecular evidence for independent evolution of photosynthetic flexibility. Plant Biology 4: 86–93.

58. WCMC-World Conservation Monitoring Centre.1992. *Global*

biodiversity: status of the Earth's living resources. London: Chapman & Hall.

59. WendtT, Canela MBF, Faria APG, Rios RI.2001. Reproductive biology and natural hybridization between two endemic species of *Pitcairnia* (Bromeliaceae). American Journal of Botany 88: 1760–1767.
60. WendtT, Canela MBF, Paz NPL, Klein DE, Rios RI.2002. Selfing facilitates reproductive isolation among three sympatric species of *Pitcairnia*. Plant Systematics and Evolution232: 201–212.
61. YoungDR, Shao G, Porter JH.1995. Spatial and temporal growth dynamics of Barrier Island shrub thickets. American Journal of Botany82: 638–645.
62. ZaluarHLT, Scarano FR.2000. Facilitação em restingas de moitas: um século de buscas por espécies focais. In: Esteves FA, Lacerda LD, eds. *Ecologia de Restingas e Lagoas Costeiras*. Rio de Janeiro: NUPEM-UFRJ, 3–23.

Chapter 2

Pollination and Reproductive Biology of Twelve Species of Neotropical Malpighiaceae: Stigma Morphology and its Implications for the Breeding System

Maria Rosângela Sigrist*[1] and
Marlies Sazima[2]

[1]Departamento de Biologia, Caixa Postal 549, Universidade Federal de Mato Grosso do Sul, 79070-900 Campo Grande, Mato Grosso do Sul, Brasil

[2]Departamento de Botânica, Caixa Postal 6109, Universidade Estadual de Campinas, 13083-970 Campinas, São Paulo, Brasil

ABSTRACT

- ***Background and Aims***: This study on reproductive biology examines the stigmatic morphology of 12 Brazilian Malpighiaceae species with regard to their pollination and breeding system.
- ***Methods***: The species were studied in natural populations of a semi-deciduous forest fragment. Style tips were processed for observation by SEM and pollen-tube growth was analyzed under fluorescence microscopy. The breeding system was investigated by isolating flowers within waterproof bags. Floral visitors were recorded through notes and photographs.
- ***Key Results***: Flowers are yellow, pink or white, protogynous, herkogamous and sometimes lack oil glands. While *Banisteriopsis pubipetala* has functional female flowers (with indehiscent anthers), 11 species present hermaphrodite flowers. Stigmas of these species may be terminal, with a slightly concave surface, or internal, consisting of a circular cavity with a large orifice, and are covered with a thin, impermeable cuticle that prevents pollen from adhering, hydrating, or germinating. Malpighiaceae have a special type of 'wet' stigma, where a secretion accumulates under the cuticle and is released by mechanical means—mainly rupture by pollinators. Even though six species show a certain degree of self-compatibility, four of them present a form of late-acting self-incompatibility, and the individual of *B. pubipetala* is agamospermous. Species of *Centris*, *Epicharis* and *Monoeca* bees pollinate these flowers, mainly collecting oil. Some *Epicharis* and *Monoeca* species collected pollen by vibration. *Paratetrapedia* and *Tetrapedia* bees are pollen and oil thieves.
- Conclusions: The Malpiguiaceae species studied are pollinator dependent, as spontaneous self pollination is limited by herkogamy, protogyny and the stigmatic cuticle. Both the oil and pollen collecting behaviours of the pollinators favour the rupture of the stigmatic cuticle and the deposition of pollen on or inside the stigmas. As fruit set rates in natural conditions are low, population fragmentation may have limited the sexual reproduction of these species.

INTRODUCTION

Malpighiaceae are predominantly tropical plants, and approx. 85 % of the currently recognized species occur in the New World (Anderson, 1979; 1990). Flowers of neotropical Malpighiaceae are relatively similar in terms of general morphology, especially as regards attraction, orientation and reward of pollinators. However, their androecia and gynoecia present considerable diversity (Anderson, 1979; 1990).

Malpighiaceae are pollinated by female Apidae bees of the tribes Centridini and Tapinotaspidini (*Monoeca* spp.), which gather the oil produced by the calyx glands, called elaiophores (Buchmann, 1987;Frankie *et al.*, 1989; Sazima and Sazima, 1989; Vogel, 1990; Pedro, 1994). Since it is used as larval provision and to waterproof the earthen brood-cell walls, this oil is very important (Neff and Simpson, 1981; Anderson, 1979; 1990; Buchmann, 1987; Sazima and Sazima, 1989; Vogel, 1988; 1990;Vinson *et al.*, 1997).

The pollination of a neotropical Malpighiaceae species was first described by Hauman-Merck (1913), who considered these calyx glands as nectaries. Later, Vogel (1974) proved that the reward produced by these glands was oil and established a new pollination system, the 'oil-flower syndrome'. This stimulated studies of the floral morphology and pollination systems of the neotropical Malpighiaceae from several ecosystems (Gottsberger, 1986; Rêgo and Albuquerque, 1989; Sazima and Sazima, 1989; Simpson, 1989; Teixeira and Machado, 2000).

Although a few genera or species of this family have had their breeding system studied (Bawa, 1974; Ruiz and Arroyo, 1978; Steiner, 1985; Barros, 1992; Teixeira and Machado, 2000), further details regarding their stigmatic morphology and its function are so far unknown. Since the morphological and physiological stigmatic features affect the reproductive biology and breeding systems (Heslop-Harrison, 2000), this information is quite important.

The purpose of the present contribution was therefore twofold: (1) to describe the reproductive biology and stigmatic morphology of four species of *Banisteriopsis*, three of *Mascagnia*, two of *Tetrapterys*, and one each of *Dicella*, *Heteropterys* and *Stigmaphyllon*; and (2) to discuss the stigma morphology of these species in the light of their pollination and breeding systems.

MATERIALS AND METHODS

Study Area and Plant Species

Field observations were made in 1994 and from 1996 to 1998 in the Santa Genebra Reserve, a 250-ha semi-deciduous forest, close to an urban area in Campinas (22°49'S, 47°7'W; altitude 670 m), São Paulo, Brazil. This forest fragment has been affected by anthropogenic disturbances (e.g. cultivation, human dwellings) since the 1950s. Included within its boundaries are areas of secondary vegetation and swamp forest. Additional information on vegetation type and climate can be found inMorellato and Leitão-Filho (1995; 1996) and Figueiredo and Sazima (2000).

The Malpighiaceae species studied are lianas commonly found at the forest edge. Their inflorescences are racemose (11 species) or cymose (*Stigmaphyllon lalandianum*), and aggregated in dichasial units. Species names and herbaria numbers are listed in the Appendix. Vouchers of these species are deposited at the herbaria of the Universidade Estadual de Campinas (UEC) and Universidade Federal de Mato Grosso do Sul (CGMS/UFMS), in Campo Grande.

General Flower Features and Stigma Morphology

Flowers and stigmas were examined *in situ* and in the laboratory. Plant material was fixed in either 70 % alcohol or FAA. Style tips were dehydrated through an ethanol series, dried by the critical-point method, mounted on stubs, and coated with gold–palladium (Dulberger *et al.*, 1994). Details of the stigmatic morphology were studied by optical, stereo and/or scanning electron microscopy.

Floral biology observations were made to verify the events of anthesis, as stigma receptivity, anther dehiscence and oil production. Stigma receptivity was checked with a magnifying glass. Pollen viability was estimated by cytoplasmic staining, using the aceto-carmine technique (Radford *et al.*, 1974).

Breeding System

The study of the breeding system of 11 species (all but *Heteropterys intermedia*) was performed by isolating flowers in waterproof bags to verify the occurrence of fertilization in spontaneous and hand self-pollination, and/or cross-pollination. Self-pollinations were made with the help of a brush. Cross-pollinated flowers were not emasculated, and pollinations were performed by rubbing anthers onto stigmas. Pollinations were made in the morning, and to verify that the stigmatic cuticle was ruptured, stigmas were examined with a magnifying glass. Flowers visited by pollinators were tagged and monitored for open pollination (control). Fruit set was recorded after full development. The number of plants per species used in each treatment is summarized in Table 1.

Table 1: Pollen viability and results of fruit set (%) in 11 species of neotropical Malpighiaceae from a semi-deciduous forest in south-eastern Brazil, after emasculation, and self-, cross- and open pollinations

Species		**Pollen viability (%)**	**Compatibility system**	**Emasculation**	**Self-pollination**		**Cross-pollination**	**Open pollination (control)**
					spontaneous	hand		
Banisteriopsis adenopoda	6	89	SC	0 (0/63)	2.5 (4/159)	24.5 (42/171)	*16.7* (25/150)	8.9 (20/225)
B. lutea	3	57	SC		0 (0/87)	5.5 (2/36)		1.8 (1/54)
B. muricata	3	84	SC		0 (0/66)	1.7 (2/117)	38.0 (16/42)	2.7 (3/108)
B. pubipetala	I	22	AG	42-5 (51/120)				18.0 (26/144)
Dicella bracteosa	12	85	SI		0 (0/42)	0 (0/45)	5.2 (2/38)	0 (0/41)
Mascagnia anisopetala	10	87	SC		1.7 (2/117)	0.7 (2/261)	3.3 (6/183)	40 (8/198)
M. cordifolia	8	77	SI			0 (0/36)	42.8 (18/42)	0 (0/57)

M. sepium	6	73	SI		0 (0/36)	0 (0/9)		11.7 (6/51)
Stigmaphyllon lalandianum	14	91	SI		0 (0/45)	0 (0/117)	7.6 (8/105)	4.9 (8/162)
Tetrapterys guilleminiana	5	74	SC	0 (0/90)	13.0 (9/69)	5.2 (3/57)		3.3 (4/120)
T. phlomoides	I	87	SC		0 (0/60)	12.9 (17/132)		5.9 (96/102)

Figures in parentheses indicate the number of samaras matured/carpels (fruits/functional carpels for *Dicella bracteosa*) used in each treatment.

n = number of individuals per species; SC = self-compatible, SI = self-incompatible, AG = agamospermous.

Pollen-tube growth was examined in at least five pistils of self-pollinated and/or cross-pollinated flowers of eight species (*Banisteriopsis adenopoda, B. lutea, B. pubipetala* and *Heteropterys intermedia* were omitted). Pistils were fixed in FAA at different intervals after pollination and analyzed under fluorescence microscopy (Martin, 1959) as for pollen-tube growth. This also established when and where any possible incompatibility reactions may occur (cf. Seavey and Bawa, 1986).

Visitors

Floral visitors were recorded over 37 days, totalling 140 h of observations. Flowers were surveyed for visitors at different times of the day, from early morning (0700 h) to late afternoon (1700 h). The behaviour of visitors was recorded through notes and photographs (cf. Sazima and Sazima, 1989). Bee specimens were collected with nets and/or waterproof bags. They were subsequently conditioned in plastic vials, identified and deposited as vouchers at the Laboratório de Abelhas of the Universidade de São Paulo.

RESULTS AND DISCUSSION

General Flower Features and Stigmatic Morphology

Flower morphology of the studied species is very similar to that of other neotropical Malpighiaceae, and readers interested in more details on the morphology, shape and variation of flower parts should refer to Anderson (1979; 1990), Gates (1982), Anderson (1997), Vogel (1990) and Barroso *et al.* (1991). Eight of these species present yellow flowers (Fig. 1A, B), while one has white or pinkish flowers and the other three have pink ones (Fig.1C). These flowers, which last 1 d (except for *B. pubipetala*, which last 2 d), range from small (<14 mm in diameter; e.g. *H. intermedia, M. anisopetala* and *T. guilleminiana*) to relatively large (18–20 mm in diameter; e.g. *Banisteriopsis* spp. and *Stigmaphyllon lalandianum*), and give off a sweet odour.

Figure 1: (A–F) Flowers of *Mascagnia sepium* and *Banisteriopsis lutea* showing the different flag petals, and the presence (A) or lack (B) of elaiophores (arrow). Scale bars = 2·5 mm. (C) Elaiophores of *M. cordifolia*with oil accumulated under the transparent cuticle (arrow). Scale bar = 1·0 mm. Arrangements of styles/stigmas and stamens in flowers of (D) *B. adenopoda*, (E) *Tetrapterys guilleminiana*, and (F) *Dicella bracteosa*. Scale bars = 5·0 mm.

The oil produced by the elaiophores accumulates under a transparent cuticle (Fig. 1C), a characteristic of most neotropical Malpighiaceae (Vogel, 1974; 1990; Simpson and Neff, 1981; Subramanian *et al.*, 1990; Cocucci*et al.*, 1996), and is the main reward for pollinators. However, *B. lutea* (Fig.1B) and some individuals of *B. muricata* and *H. intermedia* have eglandular flowers that offer only pollen as a reward (Sazima and Sazima, 1989). Eglandular flowers are found within different taxonomic levels of neotropical Malpighiaceae (Anderson, 1990) and pollen-collecting bees are the presumed pollinators (Anderson, 1979). However, Sazima and Sazima (1989) found that the eglandular morphs of *Heteropterys intermedia* and *Banisteriopsis muricata*, and even the eglandular species *Banisteriopsis lutea*, were only pollinated by oil-collecting bees, and that such morphs and species can act as deceptive blossoms within the 'oil-flower' category, and turn into 'pollen-flowers' for some bee species, thus improving their chances of pollination.

In the species studied, the flag petal, known to orientate and support the pollinators (Hauman-Merck, 1913; Anderson, 1979; Gottsberger, 1986), varies in position, form and/or limb size, length and claw thickness (Fig.1A, B). The limb can display stripes or spots of different colours (*B. lutea*, Fig. 1B) or hues (*B. adenopoda, B. muricata, M. anisopetala*; see Anderson, 1979; 1990; Lobreau-Callen, 1989).

The arrangement of stamens and pistils in the centre of the flower (Fig.1A, B, D–F) favours the contact of pollinators with both structures simultaneously during their visits. Anthers are versatile, extrorse and open longitudinally (Fig. 1D–F), usually 30 min to 1 h after floral anthesis, when stigmas are already receptive. Thus, the flowers are protogynous (except*B. pubipetala*, see below). Pollen is released near the styles, just below the stigmas (Figs 1D–F, 2A), and its viability varies between 57 and 94 % (Table 1). These species are hermaphrodite and display the 'homomorphic approach herkogamy' type; i.e. flower visitors contact the stigmas first and pollen is picked up subsequently, or when visitors depart (Webb and Lloyd, 1986; Endress, 1994), which favours cross-pollination (Webb and Lloyd, 1986). *Banisteriopsis pubipetala* has pseudo-hermaphrodite (functionally female) flowers since their anthers are indehiscent. Hermaphroditism is a more common condition in neotropical Malpighiaceae (Anderson, 1990), even though pseudo-hermaphroditism has been reported in various genera and species (Anderson, 1981; Anderson, 1982; Steiner, 1985; M.R.Sigrist, pers. obs.). Although indehiscent, the anthers of *B. pubipetala* do bear pollen but its viability is low (22 %), a feature that may be related to agamospermy (Table 1) and was also recorded for some *Peixotoa* species (Anderson, 1982).

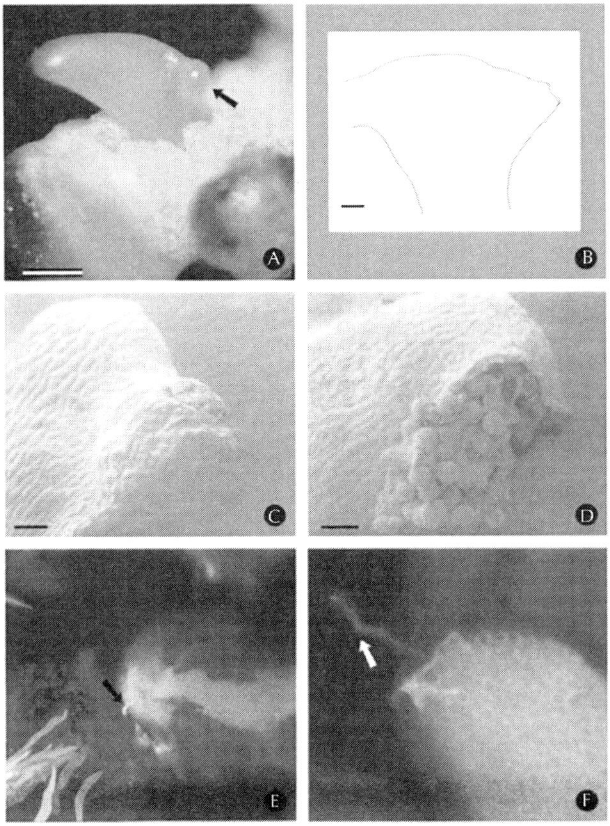

Figure 2: (A–F) Stigmas of Malpighiaceae species.*Stigmaphyllon lalandianum*: (A) secretion accumulated below the transparent stigmatic cuticle (arrow), and (B) internal stigma with an almost circular cavity and a large orifice. Scale bars = 5·0 mm and 1·0 mm, respectively. *Mascagnia cordifolia*: (C) stigma of an unpollinated flower, note the intact cuticle and the absence of pollen on it, and (D) of a pollinated flower, note the large orifice and the numerous pollen grains. Scale bars = 50 μm. Pollen tubes penetrating into ovules of *Banisteriopsis muricata* flowers 72 h after (E) self- and (F) cross-pollination (arrows). Magnification ×125 and ×312·5, respectively.

The gynoecium of the studied Malpighiaceae comprises three (eleven species) or two (*Dicella bracteosa*) superior uniovulate functional carpels, connate at base. In *D. bracteosa* one of the two ovules aborts so that only one ovule matures into a seed (W. R. Anderson, pers. comm.). Styles may be arranged in three different ways: (1) separated and disposed in a more or less broad triangle (nine species, Fig. 1D); (2) slightly separated but aligned (*H. intermedia* and *Tetrapterys guilleminiana*, Fig. 1E); or (3) grouped (*D. bracteosa*, Fig. 1F). Species of *Banisteriopsis* and *T. phlomoides* have large terminal stigmas (Fig. 1D), with a slightly concave stigmatic surface. Stigmas of *D. bracteosa* (Fig. 1F), *H. intermedia,Mascagnia* spp., *S. lalandianum* (Fig. 2A) and *T. guilleminiana* are internal, consisting of an almost circular cavity (Fig. 2B) with relatively large orifices (Fig. 2B, D). Orifices at the end of the stylar canal have also been reported in many other plant species (Heslop-Harrison and Shivanna, 1977;Dulberger *et al.*, 1994). According to Anderson (1979) and Barroso *et al.* (1991), the stigmas of neotropical Malpighiaceae vary from minute to fairly large, and from terminal to internal; when they are internal, the styles may bear apical-dorsal extensions or appendages like those of *S. lalandianum*(Fig. 2A).

In species with internal stigmas, they are orientated towards the flower centre (Fig. 1E); *D. bracteosa* was the only one to present stigmas orientated towards the sepals near the flag petal (Fig. 1F). Such stigma orientations favour deposition of a large pollen load inside the stigmatic cavity (Fig. 2D). Stigmas are covered with a thin, continuous membrane (stigmatic cuticle), which is impermeable (M. C. B. Pinheiro, pers. comm.) and protects its secretion (Fig. 2A). Pre-anthesis buds present small amounts of stigmatic secretion and their cuticle is whitish. However, in buds immediately before anthesis stigmas are receptive, swollen, with abundant secretion, and the cuticle is transparent (Fig. 2A). At that point in time, the stigmatic cuticle may function as an adaptation against excessive water loss, and temporarily protects the fluid layer below (Heslop-Harrison, 2000).

The stigmatic cuticle needs to be ruptured for pollen to adhere and hydrate, which is an essential prelude to germination (Heslop-Harrison and Heslop-Harrison, 1985). In Malpighiaceae the stigmatic cuticle is easily broken but its rupture needs mechanical factors, usually floral visitors, and mainly pollinators (Fig. 2D). The stigmas of flowers isolated in waterproof bags displayed intact cuticles and no pollen (Fig. 2C). Malpighiaceae species have a special type of 'wet' stigma (cf. Heslop-Harrison and Shivanna, 1977; Heslop-Harrison, 2000), a feature already reported for many Leguminosae (Heslop-Harrison and Heslop-Harrison, 1983; Lord and Heslop-Harrison, 1984).

Breeding System

The individual of *Banisteriopsis pubipetala* showed high fruit set in both emasculated and control flowers (Table 1), which is characteristic of agamospermous species. Marked differences in fruit set between emasculated and control flowers of this individual may be related to less intense predation of the emasculated flowers, which remained bagged for about 1 week after its manipulation, whereas control flowers were not bagged. This species is widely distributed over South America with male-sterile and male-fertile populations scattered throughout its range (W. R. Anderson, pers. comm.), and in this case agamospermy cannot be extended for the whole species. Agamospermy is uncommon in Malpighiaceae, and has been reported in only a few species, for example*Hiptage madablota* (Grant, 1981) and five species of *Peixotoa* (Anderson, 1982).

Six species showed a certain degree of self-compatibility: *Banisteriopsis adenopoda*, *B. lutea*, *B. muricata*, *Mascagnia anisopetala*, *Tetrapterys guilleminiana* and *T. phlomoides*, which had low fruit set after self-pollination (Table 1), suggesting an unimportant degree of self-compatibility. However, populations of *B. muricata* in Costa Rica (Bawa, 1974) and Venezuela (Ruiz and Arroyo, 1978) showed high levels of fruit set after manual self-pollination. Self-compatibility seems to be common in neotropical Malpighiaceae, and has already been reported for some species of *Byrsonima*, *Banisteriopsis*, *Galphimia*, *Heteropterys*, *Malpighia*and *Peixotoa* (Bawa, 1974; Barros, 1992; A. A. A. Barbosa and H. B. N. Borges, pers. comm.), and certainly in cleistogamic species of *Aspicarpa*,*Camarea*, *Gaudichaudia* and *Janusia* (Anderson, 1980). Spontaneous self-pollination was not frequent and occurred in *B. adenopoda*, *M. anisopetala*and *T. guilleminiana* (Table 1). In most of the studied Malpighiaceae, however, spontaneous self-pollination is limited by herkogamy, protogyny and the stigmatic cuticle (the latter being a mechanical system that prevents selfing), and may well be a significant determinant in the breeding behaviour of potentially self-fertile species (Heslop-Harrison and Heslop-Harrison, 1983).

Since they produced no fruits after self-pollination, *Dicella bracteosa*, *M. cordifolia*, *M. sepium* and *S. lalandianum* were considered as self-incompatible (Table 1). These species present some kind of late-acting self-incompatibility (Seavey and Bawa, 1986), since self-tubes were observed penetrating into the ovules in a similar way to cross-tubes (Fig.2E, F). This type of self-tube behaviour has been reported in self-incompatible species of other angiosperm families (Gibbs *et al.*, 1999;Gribel *et al.*, 1999). In neotropical Malpighiaceae, self-incompatibility has been reported for *Byrsonima coccolobifolia* (Barros, 1992), *B. sericea*(Teixeira and Machado, 2000), *Camarea affinis*, *Heteropterys* sp. (A. A. A. Barbosa, pers. comm.) and *Spachea membranacea* (Steiner, 1985), but the category of incompatibility has not been determined yet (cf. Seavey and Bawa, 1986).

Our results show that the majority of the species are pollinator-dependent. However, most species have low fruit set even in open (control) and/or cross-pollinations (Table 1), self-compatible species included. These results diverge from those of Bawa (1974), Ruiz and Arroyo (1978) andBarros (1992), where control and/or cross-pollination fruit-set rates were higher. The low reproductive success in the studied species suggests that inbreeding depression may be occurring in these populations (see Bawa *et al.*, 1985; Futuyma, 1993). This suggestion is supported by the fact that the study area is a relatively small forest fragment, isolated a long time ago (Morellato and Leitão-Filho, 1996).

Pollination and Visitors

Flowers of the studied species were visited by female bees of the family Apidae (Table 2). Species of *Centris, Epicharis* (Centridini) and *Monoeca*(Tapinotaspidini) pollinated the flowers, collecting mainly oil (98 % of visits) (Fig. 3A). During the visits these bees behaved as described byVogel (1974), and as further confirmed by other studies (Rêgo and Albuquerque, 1989; Sazima and Sazima, 1989; Gaglianone, 2000). The oil-collecting behaviour favours the rupture of the stigmatic cuticle and the exposure of the stigmatic surface or cavity, promoting the deposition of a large load of pollen on or inside the stigmas (Fig. 2D).

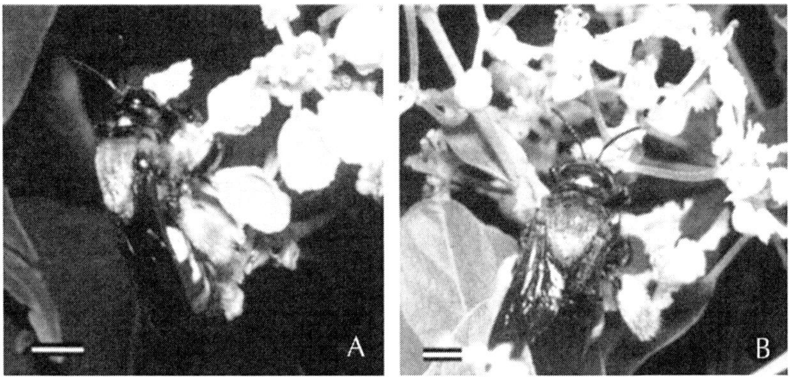

Figure 3: *Epicharis schrottkyi*pollinator of *Stigmaphyllon lalandianum* and of the eglandular morph of*Banisteriopsis muricata* while collecting (A) oil and (B) pollen by vibration. Scale bars = 5·0 mm.

Table 2: Visiting bee species and the rewards collected from the flowers of neotropical Malpighiaceae from a semi-deciduous forest in south-eastern Brazil

Bee species	Rewards	Plant species
Centris (Xanthemisia) bicolor Lepeletier	O	Blu, Dbr, Man, Sla, Tph
Centris (Melanocentris) collaris Lepeletier	O	Bad, Bmu, Dbr, Man, Sla, Tgu, Tph
Centris (Melanocentris) discolor Smith	O	Bad, Dbr, Man
Centris (Melanocentris) moc-saryi Friese	O	Bad, Bmu, Dbr, Man, Mco, Sla, Tgu
Centris (Centris) aenea Lepele-tier	O	Dbr
Centris (Paremisia) fuscata Lepeletier	O	Bad, Tgu
Centris (Heterocentris) labrosa Friese	O	Dbr
Centris (Centris) nitens Lepele-tier	O	Dbr, Mco
Centris (Hemisiella) tarsata Smith	O	Blu, Dbr, Mse

Centris (Hemisiella) trigonoides Lepeletier	O	Dbr, Sla
Epicharis (Epicharis) affnis Friese	O/P	Bad, Bmu, Dbr, Hin, Man, Sla, Tgu, Tph
Epicharis (Triepicharis) analis Lepeletier	O	Bmu, Dbr
Epicharis (Epicharis) flava Friese	O/P	all, except Mco
Epicharis (Triepicharis) schrottkyi Friese	O/P	Bad, Bmu, Dbr, Mco, Sla
Epicharis (Epicharitides) obscura Friese	O	Bad, Bmu, Sla
Monoeca sp.1	O/P	Bad, Bmu, Dbr
Monoeca sp.2	O	Bad, Man, Sla
Monoeca sp.3	O	Hin
Paratetrapedia (P.) velutina Friese	O/P	Sla, Tgu
Paratetrapedia (Lophopedia) sp.1	O/P	Bad, Dbr, Sla
Paratetrapedia (L.) sp.2	O/P	Bad, Bmu, Bpu, Dbr, Hin, Mco, Sla
Paratetrapedia (L.) sp.3	O/P	Bad
Tetrapedia (T.) diversipes Klug	O/P	Bad, Hin, Mco, Sla, Tgu

O, oil; P, pollen; Bad, *B. adenopoda*; Blu, *B. lutea*; Bmu, *B. muricata*; Bpu, *B. pubipetala*; Dbr, *D. bracteosa*; Hin, *H. intermedia*; Man, *M. anisopetala*; Mco, *M. cordifolia*; Mse, *M. sepium*; Sla, *S. lalandianum*; Tgu, *T. guilleminiana*; Tph, *T. phlomoides*.

APPENDIX

The following species were recorded and included in this study

Species	**Herbaria numbers**
Banisteriopsis adenopoda (A. Juss.) B. Gates	UEC 066311, 090721, 090722; CGMS 05986, 05994, 05995
B. lutea (Griseb.) B. Gates	UEC 090557, 090558, CGMS 05988

B. muricata (Cav.) Cuatrec.	UEC 066310, 090560; CGMS 05987
B. pubipetala (A. Juss.) Cuatrec.	UEC 090568 090570, 090708, 090709, 090710, 090726; CGMS 0985
Dicella bracteosa (A. Juss.) Griseb.	UEC 066305, 090719; CGMS 05984, 05993
Heteropterys intermedia Griseb.	UEC 066302, 090555, 090566; CGMS 05983, 05992
Mascagnia anisopetala (A. Juss.) Griseb.	UEC 066309, 090723, 090725; CGMS 05982, 05991
M. cordifolia (A. Juss.) Griseb.	UEC 066308, 090714, 090715; CGMS 05980
M. sepium (A. Juss.) Griseb.	UEC 090711, 090712, 090713; CGMS 05978
Stigmaphyllon lalandianum A. Juss.	UEC 090717, 090718; CGMS 05977, 05989, 05990
Tetrapterys guilleminiana A. Juss.	UEC 066304, 090499, 090720; CGMS 05796
T. phlomoides (Spreng.) Nied.	UEC 066306, 090559; CGMS 05979

According to Vogel (1990), Centridini bees are the main pollinators of the New World Malpighiaceae, plants that evolved under the selective impact of these oil-collecting bees. Vogel (1990) claimed that the role of the*Monoeca* species in Malpighiaceae pollination was doubtful. In the present study, however, *Monoeca* bees were pollinators of six species (Table 1); moreover *Monoeca* sp.3 was the most frequent pollinator of *H. intermedia.Monoeca* species adopt an appropriate intrafloral posture to pollinate Malpighiaceae, probably because they have tarsal combs located on the 'inner' side of their basitarsi, similarly to Centridini bees (see Neff and Simpson, 1981). Based only on putative oil-collecting organs of female bees, Neff and Simpson (1981) suggested that the *Monoeca*-Centridini lineage might represent an early split of generalized oil-collecting bees that specialized on taxa with epithelial elaiophores such as those found in the modern Malpighiaceae.

Four species of bees, *Epicharis affinis* (on *B. muricata* and *S. lalandianum*),*E. flava* (on *S. lalandianum*), *E. schrottkyi* (on *B. adenopoda, B. muricata*and *S. lalandianum*, Fig. 3B) and *Monoeca* sp.1 (on *B. adenopoda*), collected pollen by vibration (see Sazima and

Sazima, 1989). When collecting pollen, the bees adopted an intrafloral posture similar to that of oil-collecting, but for their fully extended hindlegs (Fig. 3B; see Sazima and Sazima, 1989, for behavioural details). Pollen-collecting by vibration generally occurs in flowers with poricidal anthers, although it has been recorded for many non-poricidal species of diverse angiosperm families (Vogel, 1978; Buchmann, 1985; Oliveira and Sazima, 1990). Buzz-pollination by *Epicharis* and *Centris* species has been reported for various Malpighiaceae species (Sazima and Sazima, 1989; Barros, 1992; Pedro, 1994; Gaglianone, 2000); however, this behaviour is reported here for the first time for the *Monoeca* species.

Paratetrapedia (Tapinotaspidini) and *Tetrapedia* (Tetrapedinini) bees collected oil and sometimes pollen. To collect oil, after landing, they crawled on the underside of the flower, where they scraped the oil glands. For pollen-collecting (18 % of visits), they landed on anthers and stigmas and grasped the filament or style with their mandibles. While they were collecting pollen they could rupture the stigmatic cuticle and deposit small loads of pollen. However, these bees are basically pollen and oil thieves.Vogel (1990) mentioned that *Paratetrapedia* and *Tetrapedia* species were found on Malpighiaceae as illegitimate flower visitors, but suggested that they might be legitimate pollinators of small-flowered genera.

ACKNOWLEDGEMENTS

We thank W. R. Anderson and an anonymous referee for valuable suggestions in the manuscript, H. Dolder for help in scanning electron microscopy, F. L. Modesto for help with the illustrations, I. de F. Bressan for technical assistance, M. C. H. Mamede and S. R. de M. Pedro for taxonomic identification, Núcleo de Apoio à Pesquisa Eletrônica Aplicada à Pesquisa Agropecuária (NAP-MEPA, Esalq-USP) for the use of the Zeiss DSM 940A SEM and Fundação José Pedro de Oliveira for providing permission to research in the reserve. Financial support was provided by CNPq. This contribution is part of the first author's PhD thesis.

REFERENCES

1. AndersonC.1982. A monograph of the genus *Peixotoa* (Malpighiaceae).Contributions from the University of Michigan Herbarium 15: 1–92.
2. AndersonC.1997. Monograph of *Stigmaphyllon* (Malpighiaceae). Systematic Botany 51: 1–313.
3. AndersonWR.1979. Floral conservatism in neotropical Malpighiaceae. Biotropica11: 219–223.
4. AndersonWR.1980. Cryptic self-fertilization in Malpighiaceae. Science 207: 892–893.
5. AndersonWR.1981. Malpighiaceae: the botany of Guayana Highland. Part XI.Memoirs of the New York Botanical Garden 32: 21–305.
6. AndersonWR.1990. The origin of the Malpighiaceae – the evidence from morphology. Memoirs of the New York Botanical Garden 64: 219–224.
7. BarrosMAG.1992. Fenologia da floração, estratégias reprodutivas e polinização de espécies simpátricas do gênero *Byrsonima* Rich (Malpighiaceae). Revista Brasileira de Biologia 52: 343–353.
8. BarrosoGM, Peixoto AL, Ichaso CLF, Costa CG, Guimarães, EF, Lima, HC.1991.*Sistemática de angiospermas do Brasil.* Vol. 2. Viçosa: UFV Imprensa Universitária.
9. BawaKS.1974. Breeding systems of tree species of a lowland tropical community.Evolution 28: 85–92.
10. BawaKS, Perry DR, Beach JH.1985. Reproductive biology of tropical lowland rain forest trees. I. Sexual systems and incompatibility mechanisms. American Journal of Botany 72: 331–345.
11. BuchmannSL.1985. Bees use vibration to aid pollen collecting from non-poricidal flowers. Journal of the Kansas Entomological Society 58: 517–525.
12. BuchmannSL.1987. The ecology of oil flowers and their bees. Annual Review of Ecology and Systematics 18: 343–369.
13. CocucciAA, Holgado AM, Anton AM.1996. Estudio morfológico y anatómico de los eleóforos pedicelados de *Dinemandra*

ericoides, malpigiácea endémica del desierto de Atacama, Chile. Darwiniana 34: 183–192.

14. DulbergerR, Smith MB, Bawa, KS.1994. The stigmatic orifice in *Cassia, Senna,* and *Chamaecrista* (Caesalpiniaceae): morphological variation, function during pollination, and possible adaptative significance. American Journal of Botany 81:1390–1396.
15. EndressPK.1994.*Diversity and evolutionary biology of flowers.* Cambridge: Cambridge University Press.
16. FigueiredoRA, Sazima M.2000. Pollination biology of Piperaceae species in southeastern Brazil. Annals of Botany 85: 455–460.
17. FrankieGW, Vinson SB, Williams H.1989. Ecological and evolutionary sorting of 12 sympatric species of *Centris* bees in Costa Rican dry forest. In: Bock JH, Linhart YB, eds. *The evolutionary ecology of plants*. Boulder: Westview Press, 535–549.
18. FutuymaDJ.1993.*Biologia evolutiva, 2nd edn*. Ribeirão Preto: Sociedade Brasileira de Genética/CNPq.
19. GaglianoneMC.2000. Interações de *Epicharis* (Apidae, Centridini) e flores de Malpighiaceae em µm ecossistema de cerrado. Anais do Encontro sobre Abelhas 4:246–252.
20. GatesB.1982.*Banisteriopsis, Diplopterys* (Malpighiaceae). Flora Neotropica Monograph 30: 1–237.
21. GibbsPE, Oliveira PE, Banchi MB.1999. Postzygotic control of selfing in*Hymenaea stigonocarpa* (Leguminosae – Caesalpinioideae), a bat-pollinated tree of the Brazilian cerrado. International Journal of Plant Sciences 160: 72–78.
22. GottsbergerG.1986. Some pollination strategies in neotropical savannas and forests. Plant Systematics and Evolution 152: 29–45.
23. GrantV.1981.*Plant speciation.* New York: Columbia University Press.
24. GribelR, Gibbs PE, Queiroz AL.1999. Flowering phenology and pollination biology of *Ceiba pentandra* (Bombacaeae) in Central Amazonia. Journal of Tropical Ecology 15: 247–263.

25. Hauman-MerckL.1913. Observations sur la pollination d'une Malpighiacée du genre *Stigmaphyllon*. Recherche de l'Institut Botanique de l'Université de Bruxelles9: 21–27.
26. Heslop-HarrisonJ, Heslop-Harrison Y.1983. Pollen-stigma interaction in the Leguminosae: the organization of the stigma in *Trifolium pratense* L. Annals of Botany 51: 571–583.
27. Heslop-HarrisonJ, Heslop-Harrison Y.1985. Surfaces and secretions in the pollen-stigma interaction: a brief review. Journal of Cell Science Supplement 2: 287–300.
28. Heslop-HarrisonY.2000. Control gates and micro-ecology: the pollen-stigma interaction in perspective. Annals of Botany Supplement (A) 85: 5–13.
29. Heslop-HarrisonY, Shivanna KR.1977. The receptive surface of the angiosperm stigma. Annals of Botany 41: 1233–1258.
30. Lobreau-CallenD.1989. Les Malpighiaceae et leurs pollinisateurs. Coadaptation ou coévolution? Bulletin du Museum National d'Histoire Naturelle 11: 79–94.
31. LordEM, Heslop-Harrison Y.1984. Pollen-stigma interaction in the Leguminosae: stigma organization and the breeding system in *Vicia faba* L. Annals of Botany 54:827–836.
32. MartinFW.1959. Staining and observing pollen tubes in the style by means of fluorescence. Stain Technology 34: 125–128.
33. MorellatoLPC, Leitão Filho HF.1995.*Ecologia e preservação de uma floresta tropical urbana: Reserva de Santa Genebra*. Campinas: Editora da Unicamp.
34. MorellatoLPC, Leitão Filho HF.1996. Reproductive phenology of climbers in a southeastern Brazilian forest. Biotropica 28: 180–191.
35. NeffJL, Simpson BB.1981. Oil-collecting structures in the Anthophoridae (Hymenoptera): morphology, function, and use in systematics. Journal of the Kansas Entomological Society 54: 95–123.
36. OliveiraPEAM, Sazima M.1990. Pollination biology of two species of *Kielmeyera*(Guttiferae) from Brazilian vegetation. Plant Systematics and Evolution 172: 35–49.
37. PedroSRM.1994. Interações entre abelhas e flores em uma área de cerrado no NE do estado de São Paulo: abelhas coletoras de

óleo (Hymenoptera: Apoidea: Apidae).Anais do Encontro sobre Abelhas 1: 243–255.

38. RadfordAE, Dickison WC, Massey JR, Bell CR.1974.*Vascular plant systematics*. New York: Harper & Row Publishers.
39. RêgoMAG, Albuquerque PMC.1989. Comportamento das abelhas visitantes de murici, *Byrsonima crassifolia* (L.) Kunth, Malpighiaceae. Boletim Museu Paraense Emilio Goeldi Nova Serie Zoologica 5: 179–193.
40. RuizTZ, Arroyo MTK.1978. Plant reproductive ecology of a secondary deciduous tropical forest in Venezuela. Biotropica 10: 221–230.
41. SazimaM, Sazima I.1989. Oil-gathering bees visit flowers of eglandular morphs of the oil-producing Malpighiaceae. Botanica Acta 102: 106–111.
42. SeaveySR, Bawa KS.1986. Late-acting self-incompatibility in angiosperms.Botanical Review 52: 195–219.
43. SimpsonBB.1989. Pollination biology and taxonomy of *Dinamandra* and*Dinemagonum* (Malpighiaceae). Systematic Botany 14: 408–426.
44. SimpsonBB, Neff JL.1981. Floral rewards: alternatives to pollen and nectar.Annals of the Missouri Botanical Garden 68: 301–322.
45. SteinerKE.1985. Functional dioicism in the Malpighiaceae: *Spachea membranacea* Cuatr. American Journal of Botany 72: 1537–1543.
46. SubramanianRB, Arumugasamy K, Inamdar JA.1990. Studies in the secretory gland of *Hiptage sericea* (Malpighiaceae). Nordic Journal of Botany 10: 57–62.
47. TeixeiraLAG, Machado IC.2000. Sistema de polinização e reprodução de*Byrsonima sericea* DC (Malpighiaceae). Acta Botanica Brasilica 14: 347–357.
48. VinsonSB, Williams HJ, Frankie GW, Shrum G.1997. Floral lipid chemistry of*Byrsonima crassifolia* (Malpighiaceae) and a use of floral lipids by *Centris* bees (Hymenoptera: Apidae). Biotropica 29: 76–83.
49. VogelS.1974. Ölblumen und Ölsammelnde Bienen. Tropische und Subtropische Pflanzenwelt 7: 285–547.

50. VogelS.1978. Evolutionary shifts from reward to deception in pollen flowers.Linnean Society Symposium Series 6: 89–96.
51. VogelS.1988. Die Ölblumen-Symbiosen – Parallelismus und andere Aspekte ihrer Entwicklung in Raum und Zeit. Zeitschrift fuer Zoologische Systematik und Evolutionsforschung 26: 341–362.
52. VogelS.1990. History of the Malpighiaceae in the light of pollination ecology.Memoirs of the New York Botanical Garden 55: 130–142.
53. WebbCJ, Lloyd DG.1986. The avoidance of interference between the presentation of pollen and stigmas in angiosperms. II. Herkogamy. New Zealand Journal of Botany24: 163–178.

Chapter 3

Effects of Pollination Limitation and Seed Predation on Female Reproductive Success of A Deceptive Orchid

Ryan P. Walsh, Paige M. Arnold and Helen J. Michaels

Department of Biological Sciences, Bowling Green State University, Bowling Green, OH 43402, USA

ABSTRACT

For many species of conservation significance, multiple factors limit reproduction. This research examines the contributions of plant height, number of flowers, number of stems, pollen limitation and seed predation to female reproductive success in the deceit-pollinated orchid, Cypripedium candidum. The deceptive pollination strategy employed by many orchids often results in high levels of pollen limitation. While increased floral display size may attract pollinators, C. candidum's multiple, synchronously flowering stems could promote selfing and also increase attack by weevil seed predators. To understand the joint impacts of mutualists and antagonists, we examined pollen limitation, seed predation and the effects of pollen source over two flowering seasons (2009 and 2011) in Ohio. In 2009, 36 pairs of plants size-matched by flower number, receiving either supplemental hand or open pollination, were scored for fruit maturation, mass of seeds and seed predation. Pollen supplementation increased proportion of flowers maturing into fruit, with 87 % fruit set when hand pollinated compared with 46 % for naturally pollinated flowers. Inflorescence height had a strong effect, as taller inflorescences had higher initial fruit set, while shorter stems had higher predation. Seed predation was seen in 73 % of all fruits. A parallel 2011 experiment that included a self-pollination treatment and excluded seed predators found initial and final fruit set were higher in the self and outcross pollination treatments than in the open-pollinated treatment. However, seed mass was higher in both open pollinated and outcross pollination treatments compared with hand self-pollinated. We found greater female reproductive success for taller flowering stems that simultaneously benefited from increased pollination and reduced seed predation. These studies suggest that this species is under strong reinforcing selection to increase allocation to flowering stem height. Our results may help explain the factors limiting seed production in other Cypripedium and further emphasize the importance of management in orchid conservation.

INTRODUCTION

The complex dynamics between plants, pollinators and seed predators, and how these interactions affect plant reproduction, are important in understanding the evolution of plant floral displays (Irwin and Brody 2011). While many have examined the roles of both mutualists and antagonists in plant reproduction (Gómez 2003; Strauss and Irwin 2004;Abdala-Roberts *et al.* 2009; Burkhardt *et al.* 2009; Carlson and

Holsinger 2010; Kolb and Ehrlén 2010; Ågren *et al.* 2013), relatively few have examined the effect of these interactions on the selection of floral traits (Cariveau *et al.* 2004; Parachnowitsch and Caruso 2008). Both mutualists and antagonists may forage on individuals based on floral display size, where larger floral displays provide concentrated resources for pollinators (Peakall 1989; Brody and Mitchell 1997; Mitchell *et al.* 2004) as well as seed predators (Stephens and Myers 2012). The goal of this study was to assess the relationships among mutualists, antagonists and floral display size, and how these biotic interactions influence reproductive success in a deceptive pollination system of conservation concern.

Maximizing pollen transfer efficiency has greatly shaped the evolution of a multitude of floral forms and functions in angiosperms (Barrett 2003). Increased floral display size is expected to increase pollinator attraction and visitation (Peakall 1989; Burd 1995; Aragón and Ackerman 2004;Grindeland *et al.* 2005; Li *et al.* 2011; Sletvold and Ågren 2011). In an experimental manipulation of floral display size in *Mimulus ringens*,Mitchell *et al.* (2004) found that bumblebee pollinators strongly respond to floral display size, probing more flowers in sequence on plants with large numbers of flowers. Larger floral displays also increase plant visibility, thereby increasing the attraction of pollinators from greater distances (Kindlmann and Jersáková 2006). Flowering stem height may similarly increase visibility and pollination efficiency. Synchronous flowering within a plant may also effectively increase floral display size, potentially increasing individual visibility, while synchronous flowering on the population level may increase conspecific competition for the limited pollinator resources (Crone and Lesica 2004; Crone *et al.* 2009). Measures of plant size such as numbers of stems and leaves, which may or may not correlate with floral display size, while not directly affecting the recruitment of pollinators, represent increased photosynthetic resources available for subsequent fruit maturation (Bazzaz *et al.* 1987).

Many orchids rely on a deceptive pollination strategy, a pollination strategy in which the flower provides floral cues indicating a food reward while not providing that reward (Faegri and van der Pijl 1971; Cozzolino and Widmer 2005). Deceptive pollination systems often show lower visitation and pollination relative to rewarding relatives (Nilsson 1980,1984). To account for this reduced visitation and pollination, some have hypothesized that deception reduces geitonogamous pollination by causing pollinators to flee non-rewarding patches (Smithson 2002;Johnson *et al.* 2004; Schiestl 2005; Kindlmann and Jersáková 2006; Sun *et al.* 2009), referred to as the outcrossing hypothesis (Jersáková *et al.* 2006). Orchids relying on food deception often depend on newly emergent or otherwise inexperienced insects for pollinator services (Jersáková *et al.* 2006). In *Dactylorhiza lapponica*, Sletvold *et al.* (2010) demonstrated strong pollinator-mediated selection on spur length and plant height in the open lawn community of a Norwegian fen. This and other studies of deceptive species (O'Connell and Johnston 1998; Johnson and Nilsson 1999; Gigord *et al.* 2001) indicate that a variety of floral traits (plant height, flower number, petal colour) may be targets of selection for increasing female reproductive success through pollinator visitation.

Pollen limitation, defined as the difference in seed production between open pollinated (natural pollination with no supplementation) and supplemental pollination treatments, occurs when the average open-pollinated seed production is significantly less than the average seed production of individuals receiving supplemental pollen (Knight 2003). While increased floral display size may reduce pollinator limitation, it may also substantially increase geitonogamy (de Jong *et al.* 1993; Barrett and Harder 1996; Snow *et al.* 1996; Galloway *et al.* 2002). Geitonogamy reduces female function by reducing the number or quality of the offspring, but also impacts male function by reducing the quantity of pollen available for export to other plants, also known as pollen discounting (Harder and Barrett 1995; Barrett 2003; Johnson *et al.* 2004).

After pollination, female reproductive success may be eroded by seed predation. Pre-dispersal seed predation can play an important role in determining fecundity and long-term population persistence (Louda and Potvin 1995; Russell *et al.* 2010). Chronic seed predation rates can limit population growth by reducing fecundity, while stochastic predation rates can play a more diffuse, but equally important role in population dynamics (Kolb *et al.* 2007). Just as pollinators are often attracted to large floral displays, seed predators may be attracted to the accompanying large ovule resource (Brody and Mitchell 1997; Kudoh and Whigham 1998; Galen and Cuba 2001; Irwin *et al.* 2003; Adler and Bronstein 2004; Shimamura *et al.* 2005; Stephens and Myers 2012) as many seed predators rely on ovule development to feed their offspring (Cariveau *et al.* 2004). Therefore, although dense floral resources may attract mutualist pollinators, the accompanying dense floral and fruit resources may simultaneously attract antagonist herbivores and seed predators, creating conflicting selective pressures (Louda and Potvin 1995; Strauss and Irwin 2004; Ågren *et al.* 2008). Similar to the influence of co-flowering species on plant–pollinator interactions, rates of pre-dispersal seed predation also can be influenced by community context through their attraction to other species that share seed predators (Recart *et al.* 2013).

Cypripedium species are deceptive, deciduous, terrestrial orchids with growth emerging from a subterranean rhizome (Stoutamire 1967).*Cypripedium candidum* Muhlenberg ex. Willdenow, the Small White Lady's Slipper, with yellow-green lateral sepals and petals with a white, purple-spotted labellum (Stoutamire 1967), occurs in calcareous prairies as well as fens and limestone barrens (Cusick 1980). The plants occur as single plants or large clumps (1–12 vegetative stems) containing a single large flower per vegetative stem. This floral architecture, combined with a short flowering period, severely limits the probability of reproduction by restricting opportunities for pollination. *Cypripedium candidum* flowers are pollinated by small (4–6 mm long) adrenid and halictid bees (Catling and Knerer 1980; Bowles 1983; Wake 2007). While pollen transport distances have not been specifically studied for adrenid and halictid bees, previous studies have found pollinia transport distances to vary widely (Nilsson *et al.* 1992; Kropf and Renner 2008). In the larger (8–17 mm long)*Andrena* bee species, Kropf and Renner (2008) found a maximum transport distance of 6.9 m. Data collected in a separate study (Walsh 2013) in Ohio *C. candidum* found pollen dispersal to be limited to 1 m within the focal plant.

The primary antagonist for *C. candidum*, *Stethobaris ovata* (Family: Curculiondae; subfamily: Baridinae), is a known weevil seed predator of*Cypripedium* spp. and other temperate orchid genera, with reports of adults in Canadian populations feeding on emerging shoots and flower buds (Light and MacConaill 2011). Adult weevils emerge in early spring along with *Cypripedium* shoots, and oviposit in developing fruits and, possibly stems, resulting in either fruit abortion or near total loss of the developing embryos (Light and MacConaill 2011). Predation rates on *C. parviflorum* in Canada vary from 32 to 53 % among plants in a population, depending on climate and availability of fruit resources (Light and MacConaill 2002). Little is known about the life history of the weevil, although they may complete two life cycles within a growing season (M. Light, pers. comm.).

The goal of this study was to assess the effects of plant and floral display sizes on both pollination and seed predation, and understand how these factors influence female reproductive success in the long-lived, highly specialized deceptive orchid, *C. candidum*. *Cypripedium candidum* is highly dependent on full sun in open areas and, as with many prairie species, populations begin to decline with the invasion of woody plants (Curtis 1946). In addition to the potential shading effects of encroaching woody vegetation, increased heterospecific stem density has been shown to reduce pollination and population recruitment (Wake 2007). Although the major challenges in orchid conservation research reside in understanding their symbiotic associations with fungi and improving survival following propagation (Krupnick *et al.* 2013), effective management and restoration will require a more mechanistic understanding of how habitat changes influence all biotic interactions limiting recruitment. We hypothesized that an increased floral display size would (i) increase pollinator visitation as indicated by increasing fruit set, (ii) increase geitonogamy, resulting in increased fruit abortion, decreasing fruit maturation and offspring fitness, and (iii) increase attraction of antagonists by providing an attractive resource concentration for seed predators. To address these hypotheses, we conducted two pollen limitation experiments over the course of the 2009 and 2011 field seasons in two separate sites, examining the effects of plant size on pollinator limitation and seed predation in 2009, and the effect of pollen quality in a seed predator exclusion experiment in 2011.

METHODS

The primary field site for this study was located in Northern Ohio (GPS coordinates available upon request). Historically *C. candidum* existed in at least seven Ohio counties (herbarium records OSU and BGSU). However, the Northern Ohio site is now one of the only two locations in the state where *C. candidum* remains (ODNR, pers. comm.). This site has a large, actively managed prairie area (~900 ha) with wooded areas intertwined. Vegetative cover at this location is dominated by *Andropogon gerardii,Sorghastrum nutans* and *Silphium terebinthinaceum*, with *Viola* spp.,*Sisyrinchium montanum* and *Fragaria virginiana* co-flowering with *C. candidum*. The prairie is maintained through controlled burns in early spring approximately every 3 years (J. Windus, ODNR, pers. comm.), producing large, thriving populations (total $N \sim 6000$) in the calcareous soil, although our access to the area was restricted to a subset of the total population by the Ohio Division of Natural Resources.

2009 Study

In early May 2009, we established three randomly selected 60-m line transects through a patch of *C. candidum* ($n > 250$). The population density of *C. candidum* at the prairie was previously estimated at 3.26 plants m^{-2} (range = 1–9 plants m^{-2}; SD = 2.68) (Walsh 2013). At 5-m intervals, the plant closest to a transect was selected and a second plant of equal size (number of stems and flowers) was chosen within 0.5 m of the transect on the opposite side (total $N = 72$). Any plants with flowers that had already opened, or had any pollinia removed or deposited prior to set up of the experiment were excluded from the study, resulting in the exclusion of one sample point. The standard method of assessing pollen limitation compares the female fitness of open-pollinated plants with that of plants that have had all of their flowers hand pollinated (Wesselingh 2007) to avoid redirection of resources from non-pollinated flowers, which may bias information on fruit or seed set (Knight *et al.* 2006; Wesselingh 2007). One of the paired plants was randomly chosen to receive a hand pollination treatment on all its flowers (mean = 2.06, SE = 0.14, range 1–7) with pollinia from a different population from a site at least 100 m away, while the other member of the pair was open pollinated. The numbers of flowering stems, total stems, leaves per stem and the height of each flowering stem (to the nearest 0.1 cm) were also recorded. Because the site was burned in 2009 ca. 1 month prior to sampling, surrounding vegetation was sparse during the flowering period, precluding the collection of surrounding vegetation data.

Each flower received a single pollinium from a mixed batch of pollinia gathered earlier the same day from a different population at least 100 m away. After hand pollination, flowers were bagged with a mesh (mesh opening size 3 mm × 3 mm) to prevent accidental removal of the pollinium. Mesh bags were removed after all flowers in the experiment had dehisced (ca. 2 weeks) to allow weevil predation. Capsule development was recorded 1 month after floral dehiscence (June), as well as at maturity in August when all capsules were collected and scored for insect damage and seed production. Fruit abortion was scored as the number of fruits initiated in May minus the number of fruits matured in August. We saw no evidence of browsing or herbivory other than weevil damage during the experimental period. Mature capsules were dried at 60 °C for 3 days prior to separate weighing of capsule and seed masses. Mature capsules were scored as predated when circular insect exit holes, ca. 1 mm in diameter, weevil body parts (possibly molts) and a lack of mature seeds were observed.

2011 Study

In May 2011 prior to flower opening, we set up a second pollen limitation study on a different nearby population of similar size to the 2009 study on the same property. In this experiment, a hand self-pollination treatment was added, stem and flower number were controlled and fruits were protected from weevil predation to obtain enough fruits for analysis of the effect of pollen quality on fruit set and seed mass. Five 50-m transects were randomly located across the population. For this study, only three-flowered, three-stemmed individuals were chosen to control for plant size, apply the three treatment types and to limit resource reallocation issues that might arise if non-experimental flowers set fruit. At 5-m intervals along each transect, we tagged the nearest three-flowered, three-stemmed individual and wrapped unopened flowers in a mesh to prevent visitation and weevil predation. Each stem was randomly assigned to one of three treatments: hand self-pollinated, hand-outcross pollinated (from a population >100 m away) or open pollination. Plants were checked daily for open flowers and treatments were applied when the stigma became receptive. Open-pollinated stems had mesh bags removed as soon as flowers opened, while hand-outcross and hand-selfed flowers were re-bagged after receiving appropriate treatments. Following floral dehiscence, the pollinator exclusion bags were removed. Initial fruit set was scored 2 weeks after flower dehiscence, with green, enlarged fruits scored as pollinated and pale, shrunken or missing fruits scored as a failed pollination. All fruits at this time were covered in dialysis tubing and secured at both ends to exclude insect damage. Fruit abortion was scored 4 weeks after flower dehiscence as well as at the end of the study (August 2011). We collected fruits 3 months after flower dehiscence, dried them at 65 °C for 48 h and then weighed dissected seed mass to the nearest hundred-thousandth of a gram on a Mettler AE-240 scale (Mettler-Toledo Inc.). We estimated the effect of inbreeding depression (δ) on female reproductive success by calculating the mean per-family seed production for each maternal family that matured capsules on both the self and outcross pollination treatments as $\delta = 1 - (\omega_s/\omega_x)$ where ω_s is the seed mass produced by selfing and ω_x is the seed mass produced by hand-outcross (Johnston and Schoen 1994).

Data Analysis

Analyses were performed using JMP v.9.0.2 (SAS Institute, Cary, CA, USA, 1989–2013). A generalized linear model (GLM) with an identity link function was used to assess the effects of treatment, average flowering stem height, number of flowers, number of stems and the interaction between treatment and average flowering stem height on percentage of initial capsule development (number of capsules produced/ number of flowers), proportion of capsules matured and proportion of capsules preyed upon. The number of flowers and number of stems were not significant variables in any initial models and therefore we used pooled error terms to test for other effects in the final models. The effects of number of flowers, number of stems, average height of flowering stems, treatment and percentage of capsules produced on the probability of abortion, final fruit and seed mass were analysed using a GLM. The proportions of fruit set and predation were arcsin square root transformed and all count data were log transformed. To test for proportionality of increase in response to size variables we used an ln–ln regression, testing for a slope of one (Klinkhamer and de Jong 2005; Karron and Mitchell 2012). For the 2011 study, a GLM with an identity link function examined the effect of pollination treatments on initial fruit set, final fruit set and seed mass, followed by one-way ANOVAs to test for differences among treatment groups on fruit set, abortion rates and final seed mass. *Post hoc*analyses on the one-way ANOVA's were performed using a Tukey–Kramer HSD.

RESULTS

2009 Study

Plants in the study had an average of 2.06 flowers (SE = 0.14, range 1–7) and 3.05 stems (SE = 0.23, range 1–9) per individual. Mature flowering stems had an average height of 22.6 cm (SE = 0.47), ranging between 15.1 and 40.9 cm. The average number of leaves on a plant varied little, with an average of 3.2 leaves per stem (SE = 0.01, range 3–4). Leaf length and width were not measured in this study based on prior work (Walsh 2008) that showed little between-individual variation.

In these experiments, *C. candidum* showed strong pollen limitation. Initial fruit set, measured 1 month after floral dehiscence, significantly increased in plants receiving supplemental pollen compared with open-pollinated flowers (pollination treatment $P < 0.0001$; whole model $P < 0.0001$, df = 5, AIC = 73.6, $R^2 = 0.33$; Fig. 1). Plants that received supplemental pollen had an initial fruit set of 87 % (SE = 0.037), while plants only serviced by pollinators had substantially lower initial fruit set (mean = 46 %, SE = 0.06, Fig. 1). Number of flowers and number of stems did not significantly predict the percentage of initial or final capsules produced and were omitted from further analysis. In the subsequent GLM analysis (whole model: $P < 0.0001$; df = 3; AIC = 71.2; $R^2 = 0.29$), initial fruit set was similarly affected by pollination treatment ($P < 0.0001$), but was significantly influenced by flowering stem height only for open-pollinated plants (interaction effect of treatment × the average flowering stem height, $P = 0.02$; main overall effect of height $P = 0.11$).

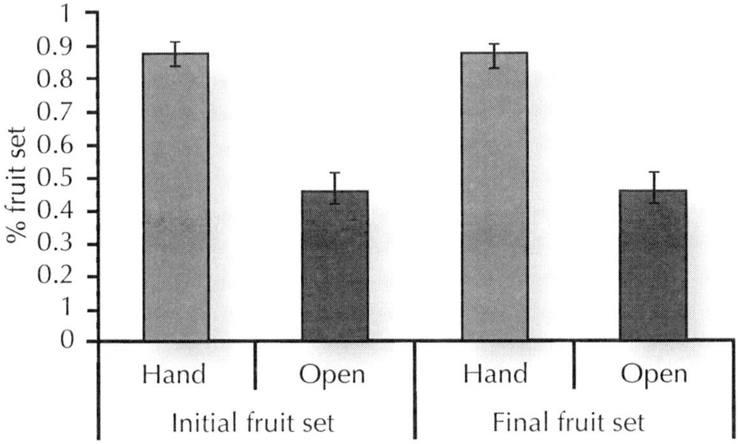

Figure 1: Effect of hand and open pollination treatments on the mean per cent of initial and final fruit set. Plants receiving supplemental pollen produced higher fruit set for both the initial time period (Hand mean = 0.87, SE = 0.037; Open mean = 0.46, SE = 0.06; $t = -4.11$, $n = 36$, $P < 0.001$) and final time period (Hand mean = 0.87, SE = 0.033; Open mean = 0.46, SE = 0.057, $t = -4.30$, $n = 36$, $P < 0.001$).

Final fruit set, measured 3 months post-floral dehiscence, also differed between pollination treatments (model $P = 0.0004$; df = 1; AIC = 121.14;$R^2 = 0.16$; Fig. 1). Study plants receiving supplemental pollen matured capsules 87 % (SE = 0.87) of the time, while only 46 % (SE = 0.057) of open-pollinated plants fully developed fruit (Fig. 1). However, no measured plant size traits (number of flowers, number of stems, height of flowering stems and number of leaves) or interaction effects explained variation in final fruit set (Fig. 2).

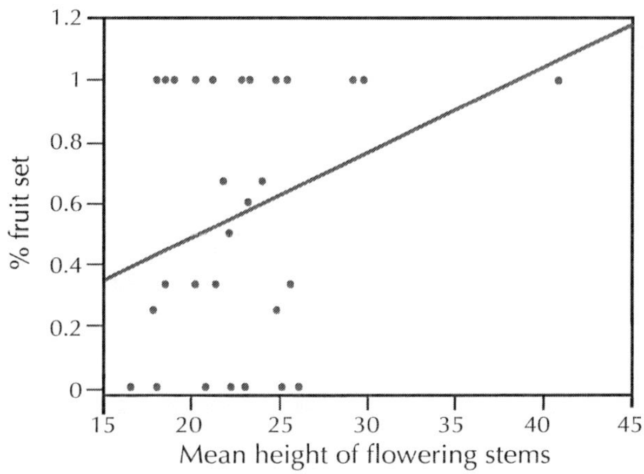

Figure 2: Relationship between per cent fruit set (number of fruits/number of flowers) of open-pollinated plants and the mean height of flowering stems. Linear regression, % fruit set = −0.34 + 0.037 × avg. height flowering stems, $F_{1,35} = 7.0$, $P = 0.012$, $R^2 = 0.17$.

Of the flowering stems that set fruit, 73 % were preyed upon. Surprisingly, predation rates were not influenced by availability of food for the weevils (numbers of flowers or fruits) or by plant size (number of stems) (whole model $P = 0.0024$; df = 5; AIC = 93.6; $R^2 = 0.23$). The average height of flowering stems was the only size variable to significantly explain the probability of predation ($P = 0.002$; whole model: $P = 0.0020$; df = 1; AIC = 98.5; $R = 0.14$), with taller stems less likely to be attacked ($F_{1,61} = 10.1$, $P = 0.002$, $R^2 = 0.14$; Fig. 3). Fruits suffering predation had poor seed production, as the seed mass of predated capsules was 89 % lower than the seed mass of capsules without predation. A total of 22 plants in the study aborted at least one fruit between initial and final fruit set measurements. Seed mass, measured as the dried and extracted seeds from the capsules, increased with increased fruit production (numbers of capsules, $P = 0.0161$) and decreased with the proportion of capsules predated ($P < 0.0001$) (whole model $P < 0.001$; df = 6; AIC = 320.5; $R^2 = 0.36$).

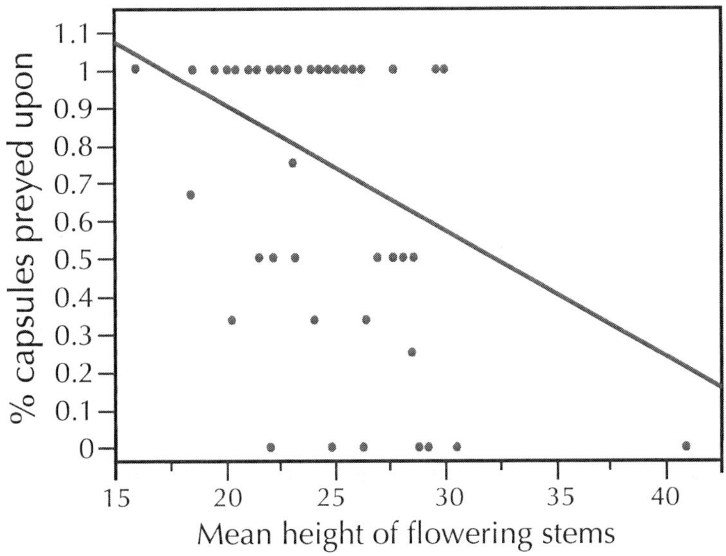

Figure 3: Relationship between the percentage of matured capsules preyed upon (number of capsules preyed upon/number of capsules produced) and the mean height of flowering stems of all plants in the study. Linear regresión, % capsules preyed upon = 1.57–0.033 × avg. height flowering stems, $F_{1,61} = 10.12$, $P = 0.023$, $R^2 = 0.144$.

2011 Study

When fruits were protected from weevil predation and plant size and flower number were controlled, initial fruit set varied with pollen source. Plants receiving hand pollination had higher fruit set 2 weeks after floral dehiscence, with plants receiving self-pollen setting 63 % of their capsules (SE = 0.08) and plants receiving outcrossed pollen setting 43 % of their capsules (SE = 0.09). In contrast, fruit set on open-pollinated plants was considerably lower, with only 16 % (SE = 0.06) of open-pollinated flowers initially setting fruit. Initial fruit set differed among pollination treatments ($F_{2,89} = 7.73$, $P = 0.0008$; Fig. 4). Plants that received hand self-pollinations had a significantly higher initial fruit set than the open-pollinated plants ($P < 0.05$, Fig. 4).

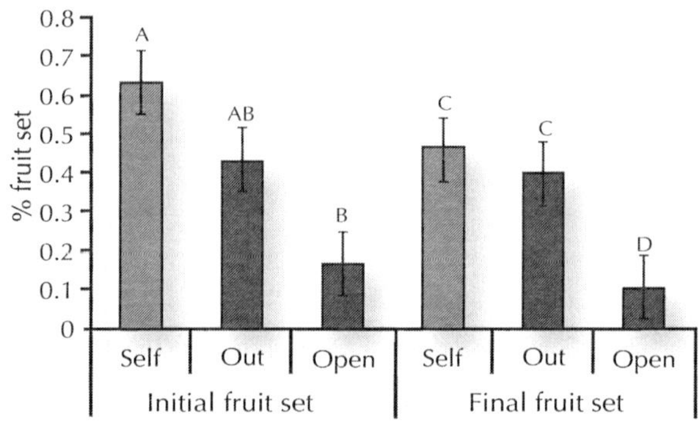

Figure 4: Effect of pollination treatments on initial and final fruit set in the 2011 study. $N = 30$. Mean fruit set for the initial (Self = 0.633, SE = 0.084; Outcross = 0.433, SE = 0.084; Open = 0.166, SE = 0.084) and final fruit maturation times (Self = 0.466, SE = 0.08; Outcross = 0.40, SE = 0.08; Open = 0.1, SE = 0.08) were significantly different as indicated by Tukey–Kramer HSD. Treatments separated by letter indicate significant difference. ANOVA: initial fruit set: $F_{2,89} = 7.73$, $P < 0.001$; final fruit set: $F_{2,89} = 5.73$, $P = 0.004$.

Final fruit set, scored 12 weeks after floral dehiscence, was affected by the pollination treatments ($F_{2,89} = 5.73$, $P = 0.0046$; Fig. 4). As in the 2009 study, plants receiving natural pollinator service were pollen limited, as self and outcross hand-pollinations matured more fruits (self mean fruits per flower = 0.46, $P = 0.0058$; outcross = 0.4, $P = 0.0292$) compared with open-pollinated stems (mean fruits per flower = 0.1; SE = 0.081). However, pollen quality did not influence the probability of fruit maturation, as the fruit set of selfed and outcrossed hand-pollination treatments were similar ($P = 0.83$; Fig. 4). However, pollen quality of the pollination treatment did significantly influence final seed mass ($P < 0.0001$, $F_{2,28} = 18.72$, Fig. 5). The open and outcross treatments produced significantly larger seed mass per capsule than the self-pollinated treatments (open = 0.027 g, SE = 0.002; outcross = 0.026 g, SE = 0.001; self = 0.017 g, SE = 0.001), while seed masses of the open pollinated and outcross hand-pollination treatments were similar (Fig. 5). There was no significant difference in abortion rates among treatments ($F_{2,89} = 1.79$, $P = 0.172$). A power analysis indicated that 81 replicates would have been needed to reach a significance level of $P < 0.05$.

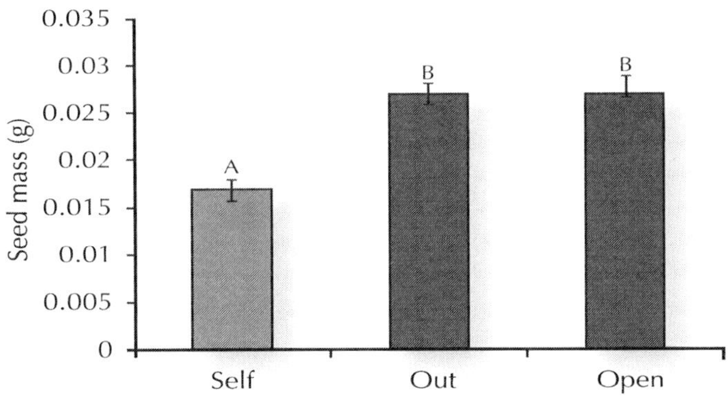

Figure 5: Effect of pollination treatment on seed mass (g) per fruit. The outcross (mean = 0.027, SE = 0.001) and open-pollinated plants (mean = 0.027, SE = 0.002) produced significantly higher seed mass than the self-pollinated plants (mean = 0.017, SE = 0.001) as indicated by Tukey–Kramer HSD, N = 29. Treatments separated by a different letter are significantly different. ANOVA: $F_{2,28} = 18.72$, $P < 0.001$, $R^2 = 0.59$.

When the 12 individuals that matured fruit on both hand-pollination treatments were used to estimate the effect of inbreeding depression on seed production, most individuals showed increased seed mass when pollinated with outcrossed pollen (mean $\delta = 0.463$; SD = 0.31). However, the effect of selfing was variable across families. Nine families had an 11–67 % reduction in seed mass when selfed, two families matured a fruit but failed to produce any selfed seed within the fruit, while one family produced similar masses of seed in both the outcross and self hand-pollinastion treatments.

DISCUSSION

Orchid species have been routinely shown to demonstrate consistent pollen limitation across multiple years (Snow and Whigham 1989;Ackerman and Montalvo 1990; Calvo 1990; Primack and Hall 1990;Dudash and Fenster 1997). Furthermore, deceptive orchids often produce only half as many fruits as their non-deceptive counterparts (Johnson and Bond 1997; Neiland and Wilcock 1998; Tremblay *et al.*

2004; Jersáková *et al.* 2006). This study provides strong evidence of pollen limitation in a deceptive orchid over two flowering seasons. In 2009, we observed a moderate fruit set from open pollination (46 %), while fruit set was greatly reduced (16.6 %) for *C. candidum* plants in 2011. This number closely parallels open-pollinated fruit set seen across multiple non-burn years at this site (16.5 %; Walsh 2013), as well as levels reported in other relatives, such as 10.5 % in *C. calceolus* (Kull 1998) and 5–13 % in *C. acaule*(O'Connell and Johnston 1998). Hand-pollinated fruit set was consistent between study years and burn/non-burn years, with at least 40 % of flowers setting fruit when supplemental pollen was provided. The mean temperature for the flowering month did not differ between 2009 and 2011 (15.5 °C); however, the 2011 study year received more than double the precipitation (18.5 cm) compared with 2009 (8.8 cm).

Pollen limitation, in principle, has two components, pollen quantity and pollen quality (Aizen and Harder 2007). In plant systems producing normal, dust-like pollen and large numbers of ovules, inadequate saturation of the stigmatic surface may result in only partial pollination. Orchids produce pollen aggregated into sac-like pollinia containing large amounts of pollen, although misplacement of the pollinia on the stigmatic surface by pollinators may result in incomplete pollination. Studies in the deceptive *Dactylorhiza* orchid have shown that flowers may need multiple visits in order to receive enough pollination for complete seed set (Sletvold and Ågren 2010; Sletvold *et al.* 2010). Although Aizen and Harder (2007)argue that pollen supplementation often involves high-quality outcross pollen that could inflate pollen deposition estimates, our 2011 study found that pollen quality manipulation in this system did not significantly increase fruit production, although it did significantly increase seed mass. Different experiments from both years show no significant change in seed mass/fruit between open-pollinated and hand-outcrossed flowers, suggesting that open-pollinated flowers are usually outcrossed.

Although numerous studies have shown that an increase in floral display and plant size increases pollen receipt and fruit maturation (Peakall 1989;Meléndez-Ackerman and Ackerman 2001; Aragón and Ackerman 2004;Mitchell *et al.* 2004; Li *et al.* 2011), we saw no effect of number of flowers, number of stems or number of leaves on pollen receipt or overall fruit maturation. The population studied was relatively dense for an orchid population, 3.26 plants m^{-2} (range = 1–9 plants

m^{-2}; SD = 2.68), and the large numbers of closely spaced individuals may have limited the ability to detect any effect of floral display size on fruiting success. However, we found strong evidence that greater flowering stem height increases initial fruit set, suggesting that taller plants were more likely to attract pollinators. Others have shown previously that the height and density of surrounding vegetation affects pollination and fruit production in deceptive orchids. Wake (2007) found increased seed set in *C. candidum*when surrounding vegetation was experimentally reduced, while the height of the flowering stem also increased pollination and fruit production in the closely related species *C. acaule* (O'Connell and Johnston 1998). Similarly, Sletvold *et al.* (2013) found strong pollinator-mediated selection for taller plants in the presence of taller vegetation in the deceptive *D. lapponica*. Given the tall grass prairie vegetation in which *C. candidum* occurs, a taller flower would be more visible to pollinators through the vegetation and therefore be more likely to receive pollinator servicing. Compared with other years (measured in a concurrent demographic survey; Walsh 2013), fruit set was unusually high in the 2009 study, which took place immediately after a controlled burn. This would suggest that both increased visibility to pollinators and increased nutrients from the burn may have contributed to this relatively high fruit set. Furthermore, taller flowering stems with increased sun exposure may offer a warmer microenvironment for the small bee pollinators in early spring, as well as greater opportunity for photosynthesis by developing fruits. Future studies on the effects of flowering stem height on seed dispersal in response to local and landscape variation in vegetation density and height may provide additional insights into the functional role of selection on plant traits affecting floral displays.

The presence of a deceptive pollination system may explain why our results with *C. candidum* are contrary to reports in the literature involving non-orchids. While other plants with larger floral displays attract pollinators from greater distances (Sih and Baltus 1987; Hessing 1988), the absence of a reward may discourage further foraging on the same plant, limiting any increase in fitness that would otherwise occur in a large, multi-flowered rewarding plant. Jersáková *et al.* (2006) cite numerous examples of deceptive orchids with reduced geitonogamy, while nectar addition experiments in deceptive orchids have found dramatic increases in self-pollination when reward is added (Johnson *et al.* 2004;Jersáková *et al.* 2006; Walsh 2013). In a deceptive system, a taller stem may increase the probability of pollinator attraction, but the visitor is expected to quickly depart after receiving no compensation for its efforts. Although food deceptive systems may increase pollen limitation compared with rewarding ones, it may be more advantageous to produce fewer, but higher quality fruits than producing additional lower quality (selfed) offspring, as seen in our 2011 data (in which seed mass substantially decreased with hand self-pollination) and several others (e.g. Tremblay *et al.* 2004; Jersáková *et al.* 2006).

These data describe a potential mechanism driving the classical outcrossing hypothesis, which explains the benefits of deceptive orchid pollination via increased outcrossing (Dafni and Ivri 1979; Nilsson 1983;Ackerman 1986; Johnson and Nilsson 1999; Jersáková *et al.* 2006). In their review of published estimates of inbreeding depression in orchids,Sletvold *et al.* (2012) noted that mean inbreeding depression for seed production was 33 %, regardless of mating system. While most individuals in this study produced greater seed mass on average when receiving outcrossed pollen (leading to a mean inbreeding depression of $\delta = 0.46$), this outcome was variable across all families, with one individual producing equal seed mass in both self and outcross treatments. The average number of flowers per plant within this population over a 4-year observation period was 1.75, with ~51 % of flowers setting fruit (Walsh 2013), indicating that although inbreeding depression in *C. candidum*might be overcome by setting an additional fruit, the floral display architecture (single flower/stem) of *C. candidum* makes this highly unlikely. Furthermore, a true estimate of inbreeding depression would require data on the germination and future growth and reproduction of the offspring and is likely to depend on environmental conditions (Cheptou and Donohue 2011; Murren and Dudash 2012). Although seed packets have been previously used to quantify germination in the field in some orchids (Rasmussen and Whigham 1993; Sletvold *et al.* 2012), attempts to germinate *C. candidum* using this method produced no seedlings over a 2-year study period (Walsh 2013).

In this study seed predators preferentially preyed on fruit with shorter flowering stems, exerting a strong concordant selective pressure reinforcing that of the pollinators. In total, seed predation heavily reduced total reproductive output of the population, with 73 % of all capsule-bearing stems attacked by a seed predator. All capsules appeared to be damaged by the same insect, most likely the weevil in the *Stethobaris*genus previously reported to prey upon *Cypripedium* fruit (Light and MacConaill 2011). Weevil predators in this genus are known to feed upon the leaves, flowers and developing capsules of many orchids, destroying most seeds by ovipositing in the maturing capsules. Contrary to our predictions, neither predation rates nor pollination success was related to other size variables such as numbers of flowers, stems and leaves. Our data suggest these weevil predators may prefer to forage on resources closer to the ground, where a lower predation risk may be associated with cover from litter and canopy vegetation, or where less energy is required to climb shorter stems to reach capsules for oviposition. Although our analyses do not indicate any response to food resource abundance such as the number of fruits on a plant, Recart *et al.* (2013) found a significant increase in abundance of another orchid weevil, *Sethobaris polita*, along with increased floral damage and reduced fruit set on a native Puerto Rican orchid, *Bletia patula*, in sites where an invasive orchid *Spathoglottis plicata* co-occurs. Complex plant–pollinator–seed predator interactions have been documented by others (Strauss and Irwin 2004), although instances of conflicting pressures seem to greatly outnumber instances of concurrent pressure.

CONCLUSIONS

Our study quantifies a three-way interaction between plants, pollinators and seed predators in a deceptive orchid system in which mutualists and antagonists are exerting concordant, reinforcing selection on a plant trait, reproductive stem height. Pollinators visited plants with taller stems more often, while another trait often associated with increasing pollinator visitation, floral display size, had no effect on pollination. Furthermore, measures of plant size, such as numbers of stems and leaves that might influence resources available for fruit maturation, did not affect female reproductive success. Seed predators may be attracted to more easily reached resources that are sheltered by surrounding vegetation and less apparent to their invertebrate predators (Marquis 1992; Palo *et al.* 1993). Additionally, the greater seed mass from both outcrossed hand and open pollination events compared with hand self-pollinations suggests that deceit pollination effectively prevents geitonogamy, so that most pollinations in *C. candidum* arise predominately from outcrossing. Although variable in magnitude and based on a small sample, inbreeding depression reduced seed mass by 11–67 % in all but one family, suggesting that conservation of small, at-risk populations that may be vulnerable to decreased pollination opportunities and increased geitonogamy should focus on facilitating outcrossing to increase recruitment.

In addition to the strong directional selective pressure on flowering stem height from both antagonists and mutualists, the height and density of the surrounding heterospecific vegetation matrix likely enhances this selection for taller flowering stems. The evolution of this complex interaction may hold an important lesson in the conservation of *Cypripedium* spp. and other deceptive plants. Management activity that controls surrounding heterospecific vegetation density and height during the flowering period may increase pollination success and fruit maturation, functioning as a cost-effective method to potentially increase rare plant offspring recruitment by modifying the pre-existing natural selective pressures on the biotic interactions of the system. However, while there have been several studies examining the effect of nectar addition on deceptive orchids (e.g. Johnson and Nilsson 1999; Johnson *et al.* 2004; Jersáková and Johnson 2006), there has been no study to date explicitly examining the demographic consequences of the deceptive pollination strategy in any plant. Although further research across multiple years and populations is needed to quantify the relative importance of pollination limitation and seed predation, orchid restoration efforts may benefit from research on whether management can be specifically targeted to ameliorate chronic pollination limitation or seed predation in orchids. As recently argued bySletvold and Ågren (2014), spatial and temporal variation in selection mediated by the biotic environment strongly affects the extent of pollinator-mediated selection. We propose that the surrounding vegetation context that provides the arena for plant–pollinator–predator interactions, as well as seed dispersal, is an important, relatively understudied component that should be considered for both managers of species of conservation concern as well as biologists seeking greater understanding of the evolution and functional significance of floral traits in these complex interactions.

SOURCES OF FUNDING

Our work was funded in part by an undergraduate research grant through the Science, Engineering and Technology Gateway Ohio (SETGO)-NSF program.

ACKNOWLEDGEMENTS

We thank the Ohio Department of Natural Resources for access to the field site, and particularly Jennifer Windus for streamlining of the permit process and providing valuable data on the *C. candidum* population. We also thank the members of the Michaels lab for their input and development of this research as well as Randy Mitchell, Karen Root, Moira van Staaden and Timothy Murnen for serving on the dissertation committee and providing critical feedback. We are indebted to two anonymous reviewers for their insights and excellent suggestions for improvements in the manuscript.

REFERENCES

1. Abdala-Roberts L, Parra-Tabla V, Salinas-Peba L, Herrera CM. Noncorrelated effects of seed predation and pollination on the perennial herb *Ruellia nudiflora*remain spatially consistent. Biological Journal of the Linnean Society 2009;96:800-807.
2. Ackerman JD. Mechanisms and evolution of food-deceptive pollination systems in Orchids. Lindleyana 1986;1:108-113.
3. Ackerman JD, Montalvo AM. Short- and long-term limitations to fruit production in a tropical Orchid. Ecology 1990;71:263-272.
4. Adler LS, Bronstein JL. Attracting antagonists: does floral nectar increase leaf herbivory? Ecology 2004;85:1519-1526.
5. Ågren J, Ehrlén J, Solbreck C. Spatio-temporal variation in fruit production and seed predation in a perennial herb influenced by habitat quality and population size.Journal of Ecology 2008;96:334-345.
6. Ågren J, Hellström F, Toräng P, Ehrlén J. Mutualists and antagonists drive among-population variation in selection and evolution of

floral display in a perennial herb.Proceedings of the National Academy of Sciences of the USA 2013;110:18202-18207.

7. Aizen MA, Harder LD. Expanding the limits of the pollen-limitation concept: effects of pollen quantity and quality. Ecology 2007;88:271-281.
8. Aragón S, Ackerman JD. Does flower color variation matter in deception pollinated *Psychilis monensis* (Orchidaceae)? Oecologia 2004;138:405-413.
9. Barrett SCH. Mating strategies in flowering plants: the outcrossing–selfing paradigm and beyond. Philosophical Transactions of the Royal Society B: Biological Sciences 2003;358:991-1004.
10. Barrett SCH, Harder LD. Ecology and evolution of plant mating. Trends in Ecology and Evolution 1996;11:73-79.
11. Bazzaz FA, Chiariello NR, Coley PD, Pitelka LF. Allocating resources to reproduction and defense. BioScience 1987;37:58-67.
12. Bowles ML. The tallgrass prairie orchids *Platanthera leucophaea* (Nutt.) Lindl. and*Cypripedium candidum* Muhl. ex Willd.: some aspects of their status, biology, and ecology, and implications toward management. Natural Areas Journal 1983;3:14-37.
13. Brody AK, Mitchell RJ. Effects of experimental manipulation of inflorescence size on pollination and pre-dispersal seed predation in the hummingbird pollinated plant*Ipomopsis aggregate*. Oecologia 1997;110:86-93.
14. Burd M. Pollinator behavioural responses to reward size in *Lobellia deckenii*: no escape from pollen limitation of seed set. Journal of Ecology 1995;83:865-872.
15. Burkhardt A, Delph LF, Bernasconi G. Benefits and costs to pollinating, seed-eating insects: the effect of flower size and fruit abortion on larval performance.Oecologia 2009;161:87-98.
16. Calvo RN. Inflorescence size and fruit distribution among individuals in three Orchid species. American Journal of Botany 1990;77:1378-1381.
17. Cariveau D, Irwin RE, Brody AK, Garcia-Mayeya LS, von der Ohe A. Direct and indirect effects of pollinators and seed predators to selection on plant and floral traits. Oikos 2004;104:15-26.

18. Carlson JE, Holsinger KE. Natural selection on inflorescence color polymorphisms in wild Protea populations: the role of pollinators, seed predators, and intertrait correlations. American Journal of Botany 2010;97:934-944.
19. Catling PM, Knerer G. Pollination of the small white lady's-slipper (*Cypripedium candidum*) in Lambton County, southern Ontario. The Canadian Field-Naturalist1980;94:435-438.
20. Cheptou PO, Donohue K. Environment-dependent inbreeding depression: its ecological and evolutionary significance. New Phytologist 2011;189:395-407.
21. Cozzolino S, Widmer A. Orchid diversity: an evolutionary consequence of deception? Trends in Ecology and Evolution 2005;20:487-494.
22. Crone EE, Lesica P. Causes of synchronous flowering in *Astragalus scaphoides*, an iteroparous perennial plant. Ecology 2004;85:1944-1954.
23. Crone EE, Miller E, Sala M. How do plants know when other plants are flowering? Resource depletion, pollen limitation and mast-seeding in a perennial wildflower.Ecology Letters 2009;12:1119-1126.
24. Curtis JT. Use of mowing in management of white ladyslippers. Journal of Wildlife Management 1946;10:303-306.
25. Cusick AW. 1980. *Cypripedium candidum* Muhl ex Willd. Ohio Heritage Database.
26. Dafni A, Ivri Y. Pollination ecology of, and hybridization between, *Orchis coriophora* L. and *O. collina* Sol. ex Russ. (Orchidaceae) in Israel. New Phytologist1979;83:181-187.
27. De Jong TJ, Waser NM, Klinkhamer PGL. Geitonogamy: the neglected side of selfing. Trends in Ecology and Evolution 1993;8:321-325.
28. Dudash MR, Fenster CB. Multiyear study of pollen limitation and cost of reproduction in the iteroparous *Silene virginica*. Ecology 1997;78:484-493.
29. Faegri K, van der Pijl L. The principles of pollination ecology. Oxford, NY:Pergamon Press; 1971.

30. Galen C, Cuba J. Down the tube: pollinators, predators, and the evolution of flower shape in the alpine skypilot, *Polemonium viscosum*. Evolution 2001;55:1963-1971.
31. Galloway LF, Cirigliano T, Gremski K. The contribution of display size and dichogamy to potential geitonogamy in *Campanula americana*. International Journal of Plant Sciences 2002;163:133-139.
32. Gigord LDB, Macnair MR, Smithson A. Negative frequency dependent selection maintains a dramatic flower color polymorphism in the rewardless orchid*Dactylorhiza sambucina* (L.) Soo. Proceedings of the National Academy of Sciences of the USA 2001;98:6253-6255.
33. Gómez J. Herbivory reduces the strength of pollinator-mediated selection in the Mediterranean herb *Erysimum mediohispanicum*: consequences for plant specialization. The American Naturalist 2003;162:242-256.
34. Grindeland JM, Sletvold N, Ims RA. Effects of floral display size and plant density on pollinator visitation rate in a natural population of *Digitalis purpurea*. Functional Ecology 2005;19:383-390.
35. Harder LD, Barrett SCH. Mating cost of large floral display in hermaphrodite plants. Nature 1995;373:512-515.
36. Hessing MB. Geitonogamous pollination and its consequences in *Gernium caespitosum*. American Journal of Botany 1988;75:1324-1333.
37. Irwin RE, Brody AK. Additive effects of herbivory, nectar robbing and seed predation on male and female fitness estimates of the host plant *Ipomopsis aggregata*. Oecologia 2011;166:681-692.
38. Irwin RE, Strauss SY, Storz S, Emerson A, Guibert G. The role of herbivores in the maintenance of a flower color polymorphism in wild radish. Ecology 2003;84:1733-1743.
39. Jersáková J, Johnson SD. Lack of floral nectar reduces self-pollination in a fly-pollinated orchid. Oecologia 2006;147:60-68.
40. Jersáková J, Johnson SD, Kindlmann P. Mechanisms and evolution of deceptive pollination in orchids. Biological Reviews of the Cambridge Philosophical Society2006;81:219-235.

41. Johnson SD, Bond WJ. Evidence for widespread pollen limitation of fruiting success in Cape wildflowers. Oecologia 1997;109:530-534.

42. Johnson SD, Nilsson LA. Pollen carryover, geitonogamy, and the evolution of deceptive pollination systems in orchids. Ecology 1999;80:2607-2619.

43. Johnson SD, Peter CI, Agren J. The effects of nectar addition on pollen removal and geitonogamy in the non-rewarding orchid *Anacamptis morio*. Proceedings of the Royal Society B: Biological Sciences 2004;271:803-809.

44. Johnston MO, Schoen DJ. On the measurement of inbreeding depression.Evolution 1994;48:1735-1741.

45. Karron JD, Mitchell RJ. Effects of floral display size on male and female reproductive success in *Mimulus ringens*. Annals of Botany 2012;109:563-570.

46. Kindlmann P, Jersáková J. Effect of floral display on reproductive success in terrestrial orchids. Folia Geobotanica 2006;41:47-60.

47. Klinkhamer PGL, de Jong T. Evolutionary ecology of plant reproductive strategies.Cambridge, UK: Cambridge University Press; 2005.

48. Knight TM. Floral density, pollen limitation, and reproductive success in *Trillium grandiflorum*. Oecologia 2003;137:557-563.

49. Knight TM, Steets JA, Ashman T-L. A quantitative synthesis of pollen supplementation experiments highlights the contribution of resource reallocation to estimates of pollen limitation. American Journal of Botany 2006;93:271-277.

50. Kolb A, Ehrlén J. Environmental context drives seed predator-mediated selection on a floral display trait. Evolutionary Ecology 2010;24:433-445.

51. Kolb A, Ehrlen J, Eriksson O. Ecological and evolutionary consequences of spatial and temporal variation in pre-dispersal seed predation. Perspectives in Plant Ecology, Evolution and Systematics 2007;9:79-100.

52. Kropf M, Renner SS. Pollinator-mediated selfing in two deceptive orchids and a review of pollinium tracking studies addressing geitonogamy. Oecologia2008;155:497-508.

53. Krupnick GA, McCormick MK, Mirenda T, Whigham DF. The

status and future of orchid conservation in North America. Annals of the Missouri Botanical Garden2013;99:180-198.

54. Kudoh H, Whigham DF. The effect of petal size manipulation on pollinator/seed-predator mediated female reproductive success of *Hibiscus moscheutos*. Oecologia1998;117:70-79.
55. Kull T. Fruit-set and recruitment in populations of *Cypripedium calceolus* L. in Estonia. Botanical Journal of the Linnean Society 1998;126:27-38.
56. Li P, Huang BQ, Pemberton RW, Luo YB, Cheng J. Floral display influences male and female reproductive success of the deceptive orchid *Phaius delavayi*. Plant Systematics and Evolution 2011;296:21-27.
57. Light MHS, MacConaill MC. Climatic influences on flowering and fruiting of*Cypripedium parviflorum* var. *pubescens*. In: Kindlmann P, Willems JH, Whigham DH, editors. Trends and fluctuations and underlying mechanisms in terrestrial orchid populations. Leiden, The Netherlands: Backhuys Publishers; 2002. p. 85-97.
58. Light MHS, MacConaill MC. Potential impact of insect herbivores on orchid conservation. European Journal of Environmental Sciences 2011;1:115-124.
59. Louda SM, Potvin MA. Effect of inflorescence-feeding insects on the demography and lifetime of a native plant. Ecology 1995;76:229.
60. Marquis RJ. Selective impact of herbivores. In: Fritz RS, Simms EL, editors. Plant resistance to herbivores and pathogens, ecology, evolution and genetics. University of Chicago Press; 1992. p. 301-325.
61. Meléndez-Ackerman EJ, Ackerman JD. Density-dependent variation in reproductive success in a terrestrial orchid. Plant Systematics and Evolution2001;227:27-36.
62. Mitchell RJ, Karron JD, Holmquist KG, Bell JM. The influence of *Mimulus ringens*floral display size on pollinator visitation patterns. Functional Ecology 2004;18:116-124.
63. Murren CJ, Dudash MR. Variation in inbreeding depression and plasticity across native and non-native field environments. Annals of Botany 2012;109:621-632.

64. Neiland MRM, Wilcock C. Fruit set, nectar reward, and rarity in the Orchidaceae.American Journal of Botany 1998;85:1657-1671.

65. Nilsson LA. The pollination ecology of *Dactylorhiza sambucina* (Orchidaceae).Botaniska Notiser 1980;133:367-385.

66. Nilsson LA. Mimesis of bellflower (*Campanula*) by the red helleborine orchid*Cephalanthera rubra*. Nature 1983;305:799-800.

67. Nilsson LA. Anthecology of *Orchis morio* (Orchidaceae) at its outpost in the north. Nova Acta Regiae Societatis Scientiarum Upsaliensis 1984;3:167-179.

68. Nilsson LA, Rabakonandrianina E, Pettersson B. Exact tracking of pollen transfer and mating in plants. Nature 1992;360:666-668.

69. O'Connell LM, Johnston MO. Male and female pollination success in a deceptive orchid, a selection study. Ecology 1998;79:1246-1260.

70. Palo TR, Gowda J, Högberg P. Species height and root symbiosis, two factors influencing antiherbivore defense of woody plants in East African savanna. Oecologia1993;93:322-326.

71. Parachnowitsch AL, Caruso CM. Predispersal seed herbivores, not pollinators, exert selection on floral traits via female fitness. Ecology 2008;89:1802-1810.

72. Peakall R. The unique pollination of *Leporella fimbriata* (Orchidaceae): pollination by pseudocopulating male ants (*Myrmecia urens*, Formicidae). Plant Systematics and Evolution 1989;167:137-148.

73. Primack RB, Hall P. Costs of reproduction in the Pink Lady's Slipper Orchid: a four-year experimental study. The American Naturalist 1990;136:638-656.

74. Rasmussen HN, Whigham DF. Seed ecology of dust seeds in-situ; a new study technique and its application in terrestrial orchids. American Journal of Botany1993;80:1374-1378.

75. Recart W, Ackerman JD, Cuevas AA. There goes the neighborhood: apparent competition between invasive and native orchids mediated by a specialist florivorous weevil. Biological Invasions 2013;15:283-293.

76. Russell FL, Rose KE, Louda SM. Seed availability and insect

herbivory limit recruitment and adult density of native tall thistle. Ecology 2010;91:3081-3093.

77. Schiestl FP. On the success of a swindle: pollination by deception in orchids.Naturwissenschaften 2005;92:255-264.

78. Shimamura R, Kachi N, Kudoh H, Whigham DF. Visitation of a specialist pollen feeder *Althaeus hibisci* Olivier (Coleoptera: Bruchidae) to flowers of *Hibiscus moscheutos* L.(Malvaceae) 1. The Journal of the Torrey Botanical Society2005;132:197-203.

79. Sih A, Baltus MS. Patch size, pollinator behavior, and pollinator limitation in catnip. Ecology 1987;68:1679-1690.

80. Sletvold N, Ågren J. Pollinator-mediated selection on floral display and spur length in the orchid *Gymnadenia conopsea*. International Journal of Plant Sciences2010;171:999-1009.

81. Sletvold N, Ågren J. Nonadditive effects of floral display and spur length on reproductive success in a deceptive orchid. Ecology 2011;92:2167-2174.

82. Sletvold N, Ågren J. There is more to pollinator-mediated selection than pollen limitation. Evolution 2014;68:1907-1918.

83. Sletvold N, Grindeland JM, Ågren J. Pollinator-mediated selection on floral display, spur length and flowering phenology in the deceptive orchid *Dactylorhiza lapponica*. New Phytologist 2010;188:385-392.

84. Sletvold N, Grindeland JM, Zu P, Ågren J. Strong inbreeding depression and local outbreeding depression in the rewarding orchid *Gymnadenia conpsea*. Conservation Genetics 2012;13:1305-1315.

85. Sletvold N, Grindeland JM, Ågren J. Vegetation context influences the strength and targets of pollinator-mediated selection in a deceptive orchid. Ecology2013;94:1236-1242.

86. Smithson A. The consequences of rewardlessness in orchids: reward-supplementation experiments with *Anacamptis morio* (Orchidaceae). American Journal of Botany 2002;89:1579-1587.

87. Snow AA, Whigham DF. Costs of flower and fruit production in *Tipularia discolor*(Orchidaceae). Ecology 1989;70:1286-1293.

88. Snow AA, Spira TP, Simpson R, Klips RA. The ecology of geitonogamous pollination. In: Lloyd DG, Barrett SCH, editors. Floral biology. New York: Chapman & Hall; 1996. p. 191-216.

89. Stephens AEA, Myers JH. Resource concentration by insects and implications for plant populations. Journal of Ecology 2012;100:923-931.
90. Stoutamire WP. Flower biology of the Lady's-Slippers (Orchidaceae: *Cypripedium*).The Michigan Botanist 1967;6:159-175.
91. Strauss SY, Irwin RE. Ecological and evolutionary consequences of multispecies plant-animal interactions. Annual Review of Ecology, Evolution, and Systematics2004;35:435-466.
92. Sun H-Q, Cheng J, Zhang F-M, Luo Y-B, Ge S. Reproductive success of non-rewarding *Cypripedium japonicum* benefits from low spatial dispersion pattern and asynchronous flowering. Annals of Botany 2009;103:1227-1237.
93. Tremblay RL, Ackerman JD, Zimmerman JK, Calvo RN. Variation in sexual reproduction in orchids and its evolutionary consequences: a spasmodic journey to diversification. Biological Journal of the Linnean Society 2004;84:1-54.
94. Wake CM. Micro-environment conditions, mycorrhizal symbiosis, and seed germination in *Cypripedium candidum*: strategies for conservation. Lankesteriana2007;7:423-426.
95. Walsh RP. USA: Bowling Green State University; 2008. *Cypripedium* (Orchidaceae) Hybridization along a Prairie/ Woodland ecotone: evidence for hybridization using morphology and genetics. MS Thesis.
96. Walsh RP. USA: Bowling Green State University; 2013. Pollination ecology and demography of a deceptive orchid. PhD Dissertation.
97. Wesselingh RA. Pollen limitation meets resource allocation: towards a comprehensive methodology. New Phytologist 2007;174:26-37.

Chapter 4

Habitat Specificity of A Threatened and Endemic, Cliff-Dwelling Halophyte

Ana D. Caperta[1,2], M. Dalila Espírito-Santo[1,2], Vasco Silva[3], Ana Ferreira[3], Ana P. Paes[1], Ana S. Róis[1], José C. Costa[1,2] and Pedro Arsénio[1,2]

[1]Plant Diversity and Conservation Group, Centro de Botânica Aplicada à Agricultura, Instituto Superior de Agronomia, University of Lisbon, Tapada da Ajuda, 1349-017 Lisbon, Portugal

[2]Centro de Ecologia Aplicada 'Prof. Baeta Neves', Instituto Superior de Agronomia, University of Lisbon, Tapada da Ajuda, 1349-017 Lisbon, Portugal

[3]Cascais Municipal Environment Company, Complexo Multiserviços, Estrada de Manique no. 1830, Alcoitão, 2645-138 Alcabideche, Portugal

ABSTRACT

Coastal areas and other saline environments are major contributors to regional and global biodiversity patterns. In these environments, rapidly changing gradients require highly specialized plants like halophytes. In European coastal cliff-tops, rocky and sandy seashores, and saltmarshes, typical halophytes from the genus Limonium are commonly found. Among them, the aneuploid tetraploid ($2n = 4x = 35$, 36, 37) Limonium multiflorum, endemic to the west coast of Portugal, is an interesting case study for investigating the ecology and conservation of a halophyte agamospermic species. Although it is listed in the IUCN red list of threatened species, information on its population size or rarity, as well as its ecology, in some respects is still unknown. Field surveys in the largest known population were performed (Raso cape, Portugal) in order to determine habitat requirements and conservation status. A total of 88 quadrats were monitored, 43 of which contained at least one L. multiflorum individual. For each sampled quadrat, four abiotic and four biotic variables as well as two spatially derived variables were recorded. Principal component analysis and cluster analysis showed narrow habitat specificity for this species which appeared to be intolerant to competition with invasive alien plants. We conclude that in situ conservation in a local 'hotspot' of this rare and vulnerable species emerges as a priority in order to ensure that biodiversity is not lost.

INTRODUCTION

Habitat assessment is a fundamental requirement for species conservation. Attempts to set plant conservation priorities revealed the need to consider attributes such as species ecological specificity, geographical rarity and rate of threat (Domínguez Lozano *et al.* 2003;Pärtel *et al.* 2005). Furthermore, a study on plant life history traits in rare versus common *taxa* also showed that infrequent species exhibited a narrower geographical range and more habitat specialization than their common relatives (Farnsworth 2007). Thus, knowledge on habitat requirements of a rare species is crucial, in particular in the case of agamospermic species (defined by an asexual breeding system) (Domínguez Lozano *et al.* 2003), to help establish the best conservation methods. In Europe, coastal vegetation presents lower diversity values when compared with other world regions (Mucina 2013) but European dry coastal terrestrial habitats, which include maritime rocks, sea-cliffs and coastal slopes, show higher values in terms of plant diversity (van der Maarel 1993; van der Maarel and van der Maarel-Versluys 1996). Among the best represented flowering plant families of these habitats are the*Plumbaginaceae* with nearly 90 % of coastal species, including most members from the genus *Limonium* (Kubitzki 1993; van der Maarel and van der Maarel-Versluys 1996). This cosmopolitan halophytic genus comprises annual and perennial species found on cliff-tops, rocky and sandy seashores, and saltmarshes (Erben 1993; Kubitzki 1993). These habitats are an important source of endemics (van der Maarel and van der Maarel-Versluys 1996), and nearly 37 % of all typical littoral species are considered threatened either because they have an extremely local distribution or because they are in decline as a result of negative human impacts in coastal areas (van der Maarel and van der Maarel-Versluys 1996).

Knowledge of habitat requirements of threatened populations from rare species selected for conservation is crucial for assuring their viability (Simberloff 1988; Brussard 1991; Schemske *et al.* 1994; Heywood and Iriondo 2003), since the establishment and expansion of a species is dependent on growth under favourable ecological conditions (e.g.Baumberger *et al.* 2012). Increasing a species' survival prospects through reintroduction or reinforcement (increase population size and diversity) of native species (Akeroyd and Wyse Jackson 1995), and knowing ecological processes in combination with demographic and genetic processes and breeding systems is therefore essential (Godefroid *et al.* 2011). However, objective data for documenting habitat preferences of rare plant species are relatively scarce, in particular those on ocean-exposed high cliff-tops subjected to salt spray transported inland by wind (Salisbury 1952; Maun 2004; Frederiksen *et al.* 2006).

The Portuguese coast is known for the richness of its flora due to a singular biogeographic position (Braun-Blanquet *et al.* 1972; Asensi *et al.* 1993; van der Maarel and van der Maarel-Versluys 1996). About 35 % of all Portuguese Natura 2000 habitats consist of coastal habitats, including Atlantic ocean-exposed cliff-tops (Costa *et al.* 2007; Martins *et al.* 2013). These locations present the original flora and vegetation on limestone, sandstone and marly clay areas, and include the rare and endemic cliff-dwelling *Limonium multiflorum* (Erben 1978, 1993; Costa *et al.* 2000;Espírito-Santo *et al.* 2012). This species is listed in Annex II of Habitats Directive (Council Directive 92/43/EEC 1992) and in the IUCN red list of threatened species (IUCN 2013), and most of its known populations are located in Portuguese NATURA 2000 Sites of Community Importance (EC 2013). Although *L. multiflorum* is considered a Portuguese crop wild relative (Magos-Brehm *et al.* 2008) and has an assessed conservation status of 'Least Concern' in the IUCN red list (IUCN 2013), information on its ecological preferences, essential for recovery plans, especially population restoration, population augmentation or population reintroduction is lacking.

In the present study we assessed habitat requirements for *L. multiflorum*. Two main questions were addressed: (i) What are the main abiotic and biotic variables favourable for its persistence? (ii) What is the vegetation cover and respective species composition associated with its presence?

METHODS

Study Species

Limonium multiflorum ($2n = 4x = 35, 36, 37$) (Erben 1993; Róis *et al.* 2012) is a perennial species endemic to a 120-km-long shoreline stretch in western Portugal (ICNF 2013) (Fig. 1) (Espírito-Santo *et al.* 2012). It is mainly found in NATURA 2000 habitat 1240 *Vegetated sea cliffs of the Mediterranean coasts with endemic Limonium* spp. and habitat 1330*Atlantic salt meadows* (APA 2011). Along the western coast, populations are found on different cliffs separated by unsuitable habitats such as acid rock cliffs or steep slopes (mostly granite and syenite), sandy beaches or pinewoods. It occurs in small populations from <10 individuals to 1000 flowering plants on average (Caperta *et al.* 2013; Róis *et al.* 2013). Seedling emergence is mainly observed in autumn, following dispersal of seeds in summer. However, a persistent seed bank is not formed in such a thin soil and only a small amount of seeds originate seedlings (A.D.C. and A.S.R., unpubl. res.). Plants then grow as vegetative rosettes for several years on cliff-tops and in nearby rocky areas with a high degree of exposure to salt spray and salt-laden winds from the Atlantic Ocean (Caperta *et al.* 2013). Flowering mainly occurs in spring and summer (April to July), although in some years, it is also observed in November and February (A.D.C. and A.S.R., unpubl. res.).

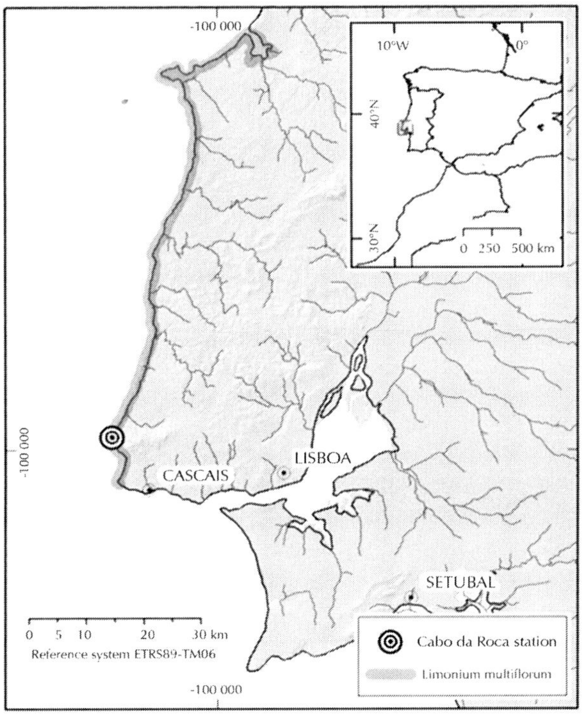

Figure 1: Map of *L. multiflorum*distribution in the western coast of Portugal.

Study Site

Our study was conducted in Raso cape, a broad promontory found in the west of Lisbon (municipality of Cascais, district of Lisbon, Portugal; mean location coordinates are 38°42'34"N and 9°29'12"W) (Fig. 2). Presently, the largest known population of *L. multiflorum* (about 1000 individuals) is found at this site, within an area of ~0.6 km^2, inside a Site of Community Importance (SCI Sintra/Cascais PTCON0008) for the Mediterranean biogeographical region. This cape consists of gently folded limestone, forming a rocky shoreline with low cliffs and numerous deep incisions along fault lines or abraded mylonitic rocks (Scheffers and Kelletat 2005). In terms of biogeographical typology this site is included in the Olissiponean District, Dividing Portuguese Sector, Sadensean-Dividing Portuguese Subprovince and Coastal Lusitan-Andalusian Province (Costa*et al.* 1998*a*; Rivas-Martínez 2007). The region's climate is well represented by the thermopluviometric diagram of Cabo da Roca station (Fig. 3), which presents a typical hyperoceanic Mediterranean climate pattern, as indicated by the low variation of average monthly temperatures and the existence of more than two consecutive dry months in the summer period. Another noteworthy feature of the region's climate is the high frequency of relatively strong winds coming from the northern and northwestern sectors (Fig. 4). Bioclimatically speaking (*sensu* Rivas-Martínez 2007), most of this region falls in the upper thermomediterranean thermotype and in the lower sub-humid to upper dry ombrotypes (Mesquita and Sousa 2009; Monteiro-Henriques 2010). Here, the predominant plant community is *Limonietum multiflori-virgati*(Costa *et al.* 1998*b*, 2012) which integrates *L. multiflorum*, *L. virgatum*,*Dactylis marina*, *Plantago coronopus* and *Crithmum maritimum*, among other species. This area encompasses different human occupations, namely buildings and other built structures like seafood tanks, off-road motorized driving and tourism which contribute to population fragmentation.

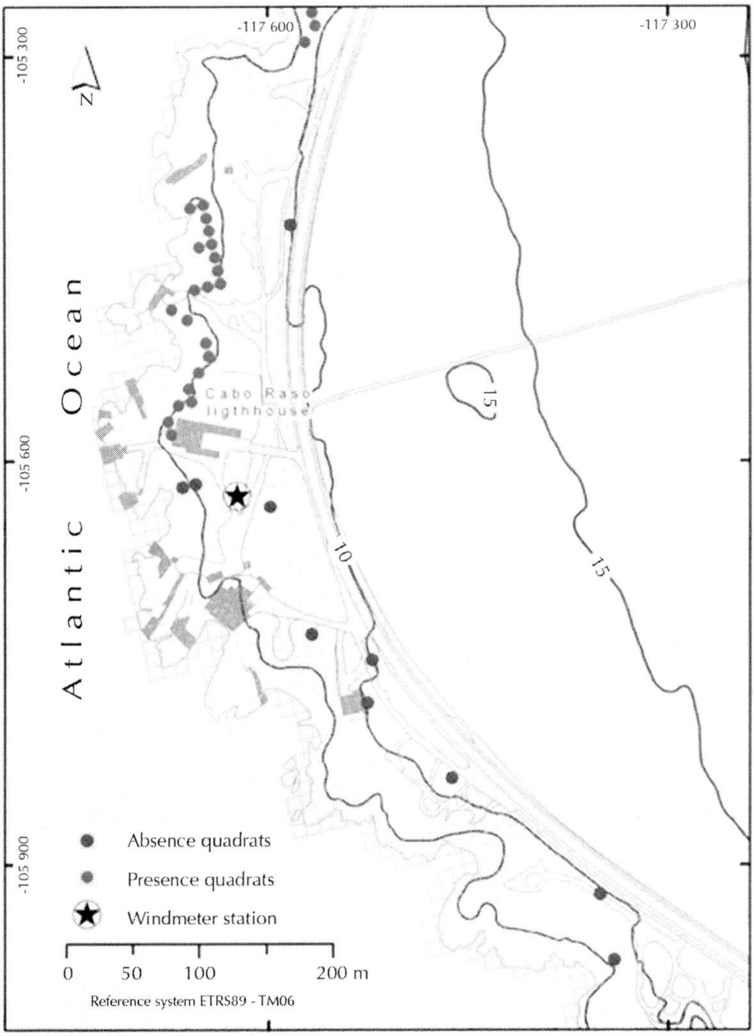

Figure 2: Schematic drawing of the quadrat sampling of *L. multiflorum* in the coast of Raso cape.

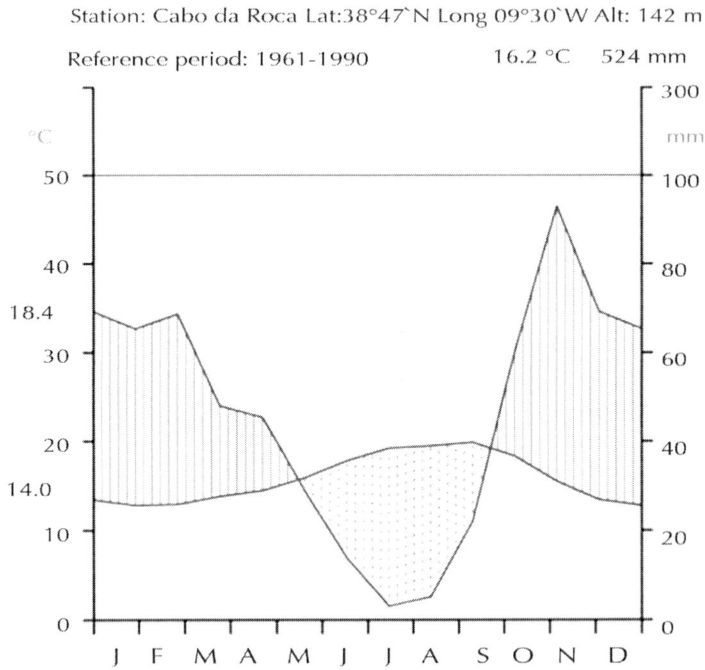

Figure 3: Thermopluviometric diagram based on the weather station in the coast of Raso cape. The red line corresponds to mean monthly temperature (°C) and the blue line corresponds to mean monthly precipitation (mm).

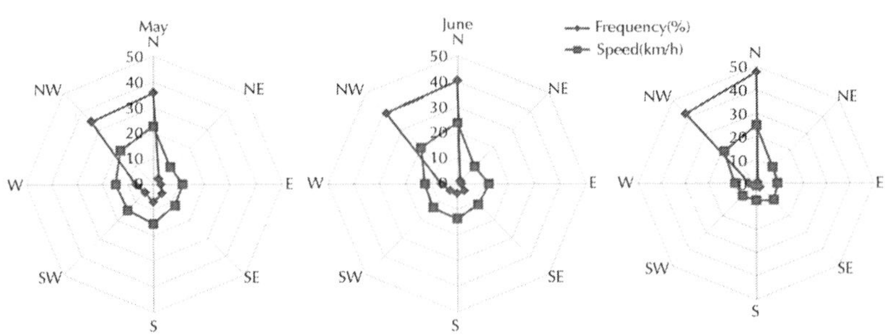

Figure 4: Wind regime diagrams based on the weather station in the coast of Raso cape during *L. multiflorum* flowering peak (May–July).

Sampling Design

Field surveys used a grid of 10 × 10 m quadrats which was overlaid over the aerial photographs of the site and later transferred to the site using global positioning system (GPS) receivers. A total of 88 quadrats were surveyed, 43 of which contained at least one *L. multiflorum* individual (presence quadrat) and the remaining 45 were randomly selected quadrats containing no individuals from this species (absence quadrat) (Fig. 2). The random selection procedure was performed using R software's 'sample' routine (R Core Team 2013). In quadrats where *L. multiflorum* was present censuses were carried out from April to June 2013. For each sampled quadrat four abiotic variables, namely rock formation (RockForm), cobble (Cobble), coarse sand (CoarseSd) and fine sand (FineSand) (all expressed in %), were recorded. Four biotic variables related to vegetation cover (Coverage) as well as invasive non-native species (INNS) cover, dead organic matter (lignified or not) (DOM) cover and litter (Residues) cover were also estimated (expressed in %) in each quadrat. To facilitate data collection, each quadrat was then sub-divided into sub-quadrats of 1 m^2, and the percent coverage value obtained for the sub-quadrats of the same 100 m^2 quadrat were averaged. Two spatial variables, distance from coastline (Dist_coa) and mean quadrat slope (Mean_slo), were also derived using the ArcGIS® 10.0 software, by ESRI.

Data Analysis

To define *L. multiflorum* habitat requirements we performed a principal component analysis (PCA) on the resulting data matrix using abiotic, biotic and spatially derived variables. In order to discriminate the vegetation structure and composition in the monitored quadrats, a cluster analysis using Euclidean distance and Ward clustering method (Ward 1963) was performed using the set of quadrats containing *L. multiflorum*. To ensure that the group of quadrats (clusters) had the same influence from the environmental variables, Kruskal–Wallis tests of variance were performed (Sokal and Rohlf 1997; Zar 2010). To complete habitat characterization of*L. multiflorum,* the most frequent and abundant plant species within each cluster was defined as described in Baumberger *et al.* (2012). The mean coverage of each species within each group was also calculated. All statistical analyses were conducted using the STATISTICA software (StatSoft V10). Bioclimatic and biogeographic nomenclature followed the proposals of Rivas-Martínez (2007). Taxonomic nomenclature followedMenezes de Sequeira *et al.* (2011).

RESULTS

Influence of Environmental Variables on the Presence/Absence of *L. multiflorum*

On the western coast most *L. multiflorum* populations occur at inaccessible cliff sites and/or show very small size (Róis *et al.* 2013). Under these circumstances, this study was conducted at the best accessible site and on the largest known population for this species, in Raso cape. Study site rocks were bare of vegetation up to ~7–8 m asl, due to strong exposure to Atlantic storms and waves over the years.

Of the 88 quadrats sampled, only 43 contained at least one *L. multiflorum*individual (presence) while control quadrats did not show any individual from this species (absence). Considering the presence quadrats, *L. multiflorum* mean coverage was very low (about 0.67 %). Only one of the variables measured in each quadrat fitted a normal distribution (variable 'Mean_slo'), and the other nine failed to do so, even after a logarithmic transformation. Therefore, the remaining analyses were performed using the original (untransformed) values. Remarkably, visual inspection of PCA revealed that in the first two represented ordinations, quadrats showing the presence of *L. multiflorum* individuals were highly correlated with rock formation (variable 'RockForm') and non-correlated with high coverage (variable 'vegetation cover') (Fig. 5). Considering the first two principal components of PCA, the percentage of accumulated variance was 64.4 % (Fig. 5). The first axis accounted for 44.8 % of the variance and was explained by vegetation cover, while the second axis accounted for 19.7 % of the variance explained by distance from coast (variable 'Dist_coa').

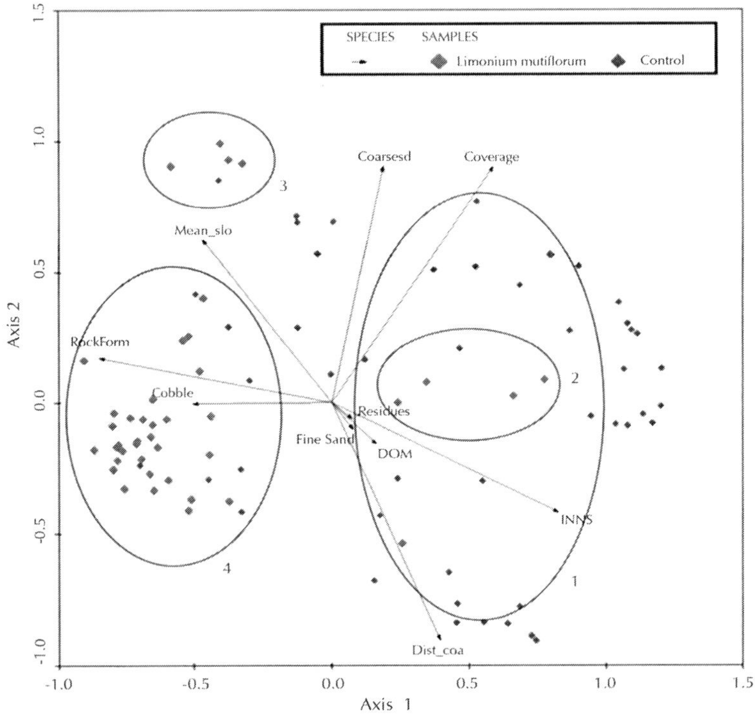

Figure 5: Two first axes of PCA based on *L. multiflorum* presence and absence quadrats. Red symbols mean that *L. multiflorum* was recorded whereas blue symbols signify that it was absent. Axis 1 represents 44.8 % of variation, and Axis 2, 19.7 %. CoarseSd, coarse sand; Cobble, cobble; Coverage, vegetation cover; Dist_coa, distance from coast; DOM, dead organic matter cover; Fine-Sand, fine sand; INNS, invasive non-native species cover; Mean_slo, mean slope; Residues, litter cover; RockForm, rock formation.

Influence of Environmental Variables in *L. multiflorum* Persistence

A cluster analysis was performed using the dataset of quadrats showing *L. multiflorum* presence. The data obtained revealed the existence of four floristic groups among the sampled quadrats (Fig. 6; Tables 1 and 2). Environmental conditions differed between Group 4 and the other three floristically defined groups. Clusters 3 and 4 were clearly individualized in PCA. The first groups (Groups 1, 2 and 3) presented high vegetation coverage percentage (variable 'coverage') combined with low rock outcrop percentages (variable 'RockForm') whereas the opposite occurred in Group 4, which showed the highest *L. multiflorum* cover percentage. Also, in this latter group low percentages of invasive non-native species (variable 'INNS') were found. Although in both Groups 1 and 2, 'INNS' percentages were greater than those in the other groups, Group 2 differed in terms of fine sand percentages (variable 'FineSand').

Table 1: Mean values ± standard error of abiotic, biotic and spatial-related variables in each of four environmental groups based on the presence data of PCA. Bold letters within a row indicate non-significant with post hoc multiple comparison Tukey›s test ($\alpha = 0.05$; $P < 0.001$). *P*value <0.001. ns, non-significant; CoarseSD, coarse sand; Cobble, cobble; Coverage, vegetation cover; Dist_coa, distance from coast; DOM, dead organic matter cover; FineSd, fine sand; INNS, invasive non-native species cover; Mean_slo, mean slope; Residues, litter cover; RockForm, rock formation.

	Environmental groups *1 (n* = 3)	**2 (n =3)**	**3 *(n* =3)**	**4 (n =34)**
Coverage (%)*	54.83± 47.54 (a)	47.17± 11.07 (a)	37.67± 11.68 (a)	3.10± 7.44 (b)
FineSd (%)*	0.00± 0.00 (a)	3.17± 2.75 (b)	0.00± 0.00 (a)	0.00± 0.00 (a)
INNS (%)*	73.00± 6.24 (a)	33.00± 3.46 (b)	0.00± 0.00 (c)	0.88± 2.55 (c)
RockForm (%)*	5.00± 4.00 (a)	6.00± 5.29 (a)	40.00± 11.79 (a, b)	74.49± 23.58 (b)
CoarseSd (%) ns	5.67± 8.14	0.00+ 0.00	11.67+ 4.16	3.37+ 9.17
Dist_coa (m) ns	11.36± 8.24	16.51+ 3.95	0.00+ 0.00	16.31+ 9.88
Cobble (%) ns	4.77± 6.60	45.00+ 8.23	8.67+ 10.017	13.63+ 19.85

DOM (%) ns	2.67± 4.62	0.00+ 0.00	5.33+ 9.24	0.41+ 1.23
Mean_slo (m) ns	16.63± 5.66	17.25± 0.82	68.92± 28.97	25.21± 21.50
Residues (%) ns	0.33± 0.58	0.42+ 0.52	0.00+ 0.00	0.09+ 0.19

Table 2: Mean coverage (%) of the most frequent species in each of the four environmental groups (clusters) based on the presence data of PCA.

Floristic list		Mean coverage (%) 1 2			
		1	2	3	4
Psammophilous	*Armeria welwitschii* Boiss.	1.17	2.34	9.67	0.03
	Inula crithmoides L	0.00	0.17	0.12	1.33
	Herniaria maritima Link	0.00	0.67	0.00	0.07
	Lobularia maritima (L.) Desv. subsp. *maritima*	0.00	0.00	0.20	0.00
	Parapholis incurva (L.) C.E.Hubb.	0.50	0.00	21.33	0.70
	Helichrysum italicum (Roth) G.Don subsp. *picardi* (Boiss. & Reut.) Franco	0.33	0.00	0.17	0.03
	Andryala arenaria (DC.) Boiss. & Reut. subsp. *arenaria*	0.17	0.33	0.58	0.06
	Otanthus maritimus (L.) Hoffmanns. & Link	0.00	2.75	0.00	0.02
	Elymus farctus (Viv.) Runemark ex Melderis subsp. *boreo-atlanticus* (Simonet & Guin.) Melderis	3.33	0.77	1.00	0.98
	Lotus creticus L.	0.00	0.01	0.25	0.40
	Medicago minima (L.) L.	0.00	050	1.00	0.03

	Crucianella maritima L.	0.13	0.00	0.00	0.03
	Cakile maritima Scop.	0.00	0.00	0.00	0.01
	Dactylis smithii Link subsp. *marina* (Borrill) Parker	0.02	0.01	0.25	0.09
Chamosphilous	*Limonium ovalifolium* (Poir.) Kuntze	0.00	0.00	1.33	0.03
	Limonium virgatum (Willd.) Fourr.	0.17	0.00	0.07	0.07
	Plantago coronopus L.	0.02	0.17	0.75	0.07
	Limonium multiflorum Erben	0.03	0.05	0.11	0.84
	Frankenia laevis L.	0.83	1.67	1.00	1.03
	Crithmum maritimum L.	0.30	3.67	0.23	0.66
	Catapodium marinum (L.) C.E.Hubb.	0.00	0.00	0.00	0.03
Nitrophilous	*Carpobrotus edulis* (L.) N.E.Br.	73.00	33.00	0.00	0.88
	Beta maritima L.	0.08	0.33	0.50	0.28
	Daucus halophilus Brot.	0.00	0.00	0.00	0.01
	Leontodon taraxacoides (Vill.) Mérat subsp. *taraxacoides*	0.00	0.00	0.00	0.01
	Parapholis filiformis (Roth) C.E.Hubb.	0.50	0.00	0.00	0.06
	Polypogon maritimus Willd.	0.00	0.00	0.00	0.02

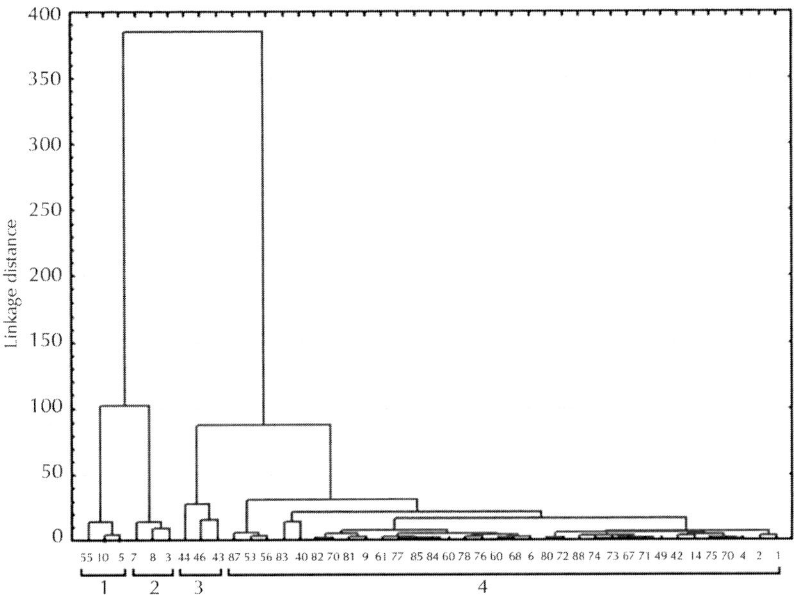

Figure 6: Cluster dendogram of the four clusters containing *L. multiflorum*. The dissimilarity between groups was calculated based on species variables using Euclidean distance and the Ward aggregation method. Cluster definition 1 accounts for a high percentage of *C. edulis* (>65 %); Cluster 2 by a moderate percentage of *C. edulis* (<35 %); Cluster 3 is defined by *A. welwitschii* (9.67 %) and *P. incurva* (21.33 %); and Cluster 4 represents quadrats with the highest *L. multiflorum* frequency (0.84 %).

Vegetation composition also varied between clusters. Cluster 1 was dominated by *Carpobrotus edulis* with a coverage percentage >65 % whereas Cluster 2 presented a moderate coverage of *C. edulis* (<35 %). In this cluster significant mean coverage was observed for psammophilous species like the perennials *Armeria welwitschii* and *Elymus farctus* subsp.*boreo-atlanticus*. As for Cluster 3, significant frequencies of the perennial*Limonium ovalifolium* and the therophyte *Parapholis incurva* were revealed. Finally, Cluster 4 which had the greatest coverage percentage of*L. multiflorum* (0.835 %) of all clusters appeared to be associated with chasmophytic species such as *C. maritimum*, and other perennials like*Frankenia laevis* and *Inula crithmoides*.

DISCUSSION

L. multiflorum Displays Narrow Habitat Specificity

In this study we show that *L. multiflorum* grows on sea-cliffs preferentially up to 50 m asl in karstic crevices within exposed rocks more or less filled with substrate, or on shallow soil above the rock strata and on scree slopes. The presence of *L. multiflorum* was favoured by the presence of rock formation, cobble, low percentage of vegetation coverage, low frequency of invasive non-native species and closest distance to coast. In fact, as pointed out by Warming (1909), *L. multiflorum*'s preference for salt-rich rock crevices seems to be a strategy to use such locations as refuges, thus avoiding competition with many other coastal plants that are unable to colonize such habitats. Furthermore, as it happens in endemic*Limonium* spp., populations tend to be fragmented and usually show low population sizes as a consequence of many social and economic activities like urbanization, tourism and traffic (Rodríguez *et al.* 2003; Palop-Esteban *et al.* 2007; Suárez-García *et al.* 2009; Khan *et al.* 2012; Róis *et al.* 2013) on their habitats. This is also the case of *L. multiflorum* (ICNF 2013), in particular, in the Raso cape population (A.F. and V.S., unpubl. res.).

Vegetation composition significantly differs among the four floristically defined groups. Cluster 1 is dominated by the alien plant *C. edulis* with *E. farctus* subsp. *boreo-atlanticus*'s due to nitrophilous coarse sand dunes corresponding to *Elytrigietum junceo-boreoatlanticae* association (Costa*et al.* 2012). The presence of *L. multiflorum* in this cluster can be explained by 5 % of rock outcrop, uncovered with sand. Cluster 2 includes the farthest quadrats from the sea where rock depressions filled up with fine sand and cobble appear. In this cluster *A. welwitschii, F. laevis* and *L. multiflorum* predominate. This cluster presents a poor floristic composition, as it happens in the previous cluster, due to invasion of *C. edulis*, possibly related with the proximity to roads (higher nitrification).Cluster 3 is typically formed by species of rocky substrate, with a high slope and direct influence of wind loaded with salt. Here *L. ovalifolium, C. maritimum, A. welwitschii, F. laevis, L. multiflorum, P. coronopus* form the perennial community *Limonietum* *multifloro-virgati* (Costa *et al.* 1998*b*,2012). In their clearings *P. incurva*, *Medicago minima*, *Andryala arenaria*form the poor, annual association *Parapholido incurvae–Catapodietum marini*. Also, the presence of *Lotus creticus* is due to coarse sand. Cluster 4 also exhibits species that constitute the *Limonietum multifloro-virgati*association (Costa *et al.* 1998*b*, 2012), namely *C. maritimum*, *L. multiflorum*, *I. crithmoides*, *F. laevis*, *P. coronopus*. The referred species are at their ecological optimum in this rocky sea-cliff cluster although a small percentage of coarse sand explains the presence of *E. farctus* subsp.*boreo-atlanticus* and *L. creticus*. Plant survival in this ecological context is difficult due to low availability of fresh water, low level of essential nutrients and the abrasive action of strong winds laden with salt and increasing conditions of extreme dryness (Costa *et al.* 1998*b*). To overcome this some plants respond with adaptations, such as adopting the hemicryptophyte or therophyte life forms, presenting succulent leaves and/or salt glands in the leaves (Warming 1909; Costa 2001; Grigore *et al.* 2014).

L. multiflorum Appears to be Intolerant to Competition with Invasive Species

Our results show that both native and non-native vegetation cover are negatively correlated with *L. multiflorum* presence, emphasizing its preferences for sites where competition with other species is low. Among the species having a greater negative effect in their persistence is the exotic, invasive, crawling *C. edulis*. This species commonly invades coastal habitats in Mediterranean Europe forming mantles on maritime rocks, cliffs and sand dunes (D'Antonio 1990; Traveset *et al.* 2008; Pyšek *et al.* 2013) competing directly with native plant species, suppressing the growth and establishment of other plants (D'Antonio and Mahall 1991; Vilà*et al.* 2006), altering certain soil parameters (Vilà *et al.* 2006; Conser and Connor 2009) or creating litter accumulation on the soil surface (Molinari*et al.* 2007). Although *C. edulis* seems to develop well on rocky substrates at sites where seagulls' (*Larus* sp.) nutrient-rich droppings are frequent (M.D.E.-S., unpubl. res.), in our study it seems to compete mainly with species showing preference for sand substrates. Remarkably, some littoral species show strong resilience to *C. edulis* invasion, like the narrow endemic cliff-species *Armeria pseudoarmeria* (Pinto-Ricardo *et al.* 2010) or other *Limonium* spp. (Snogerup 1971; Suehs *et al.* 2003; Campos *et al.* 2004).

Conservation Issues

For the effective protection of the rare and endemic *L. multiflorum* not only ecological data but also life story, demography and genetic data should be taken into account to ensure that maximum genetic variation is preserved. In this context, one of the factors to be considered in conserving *L. multiflorum* is related with chromosome polymorphisms since euploid and aneuploid cytotypes occur within populations (Róis *et al.* 2012). Although, it was first described as an unbalanced aneuploid tetraploid ($2n = 4x = 35$) (Erben 1978), both unbalanced and balanced tetraploid cytotypes ($2n = 4x = 35, 36, 37$) and diverse karyological polymorphisms were found within and among populations, with greater variability in the Raso cape population (Róis *et al.* 2012). This has also been observed in other *Limonium* taxa (e.g. Castro and Rosseló 2007), and even in a variety of genera from different plant families (e.g. *Brassicaceae*,Böcher 1954; *Campanulaceae*, James 1965; *Onagraceae*, Bloom 1974;*Plumbaginaceae*, Erben 1978; *Poaceae*, Sieber and Murray 1981;*Ranunculaceae*, Borchers-Kolb 1983). These cytological differences should be considered to ensure that maximum genetic variation is preserved. Furthermore, the resultant combination of incompatible cytotypes could result in the failure of reintroduced plants to reproduce and may bring reproductive instability to augmented populations (Young and Murray 2000; Ennos *et al.* 2005; Severns and Liston 2008).

Noticeably, male sterility and gynodioecism are found in plants from experimental collections (Róis *et al.* 2012) and field plants (Heike Sprenger and A.D.C., unpubl. res.). However, these plants are capable of producing large numbers of viable seeds with variable ploidy levels, as revealed by flow cytometry seed screening (Róis *et al.* 2012) and reproduce through apomixis (A.S.R. and A.D.C., unpubl. res.). This form of uniparental reproduction is found in 'taxonomically complex groups' like in other*Limonium* spp. (Erben 1978; Palacios *et al.* 2000; Lledó *et al.* 2005) and also in dandelions (*Taraxacum* spp.; van Dijk 2003), blackberries (*Rubus fruticosus* agg.; Salvini *et al.* 2006) and in *Ranunculus* spp. (Hörandl *et al.* 2009), which generates a diverse mixture of related individuals (Ennos *et al.* 2005).

While some *L. multiflorum* populations are located in natural parks (Parque Natural Sintra-Cascais) and other areas under legal protection (e.g. Site of Community Importance—SCI Sintra/Cascais PTCON0008; SCI Peniche/Sta Cruz PTCON0056), in the past decades no information on population size along its distribution range has been available. However, we hypothesized that population decline might have occurred due to increasing urban development in the coastal areas and the impact of tourism. For instance, in Raso cape, sightseeing tourists, sports fishing and other socio-economic activities, together with competition associated with invasion by non-native species are cumulatively leading to population decline (V.S., A.F. and A.D.C., unpubl. res.). Notably, methylation-sensitive amplified polymorphism (MSAP) data from *L. multiflorum* natural populations reveal low levels of genetic/epigenetic diversity (Róis *et al.* 2013). Altogether these data mean that none of the populations analysed could restore the genotypic diversity observed in any other *L. multiflorum* population. Hence, habitat protection emerges as the top priority to prevent population extinction of the narrow-specialist *L. multiflorum*. For the conservation of this species, management actions such as those referred by Laguna *et al.* (2013), namely limiting access to populations, for example, the experimental exclusion of herbivores by means of fences to protect populations of *L. dufourii* in 'Marsh dels Moros' (Spain) and to perform selective removal and control of invasive species (Laguna *et al.* 2013), are required. Other technical solutions could be population reinforcements, reintroduction or establishment of new populations, the goal of which is the introduction of new specimens. In the case of *L. multiflorum*, attempts to create new populations through experimental actions like sowing seeds directly *in situ* did not prove to be successful (A.D.C. and V.S., unpubl. res.) but population reinforcements by planting newly produced specimens using seed stock that originated from the same natural populations (e.g. Raso cape) appears to give better results (Capertaet *al.* 2013). Complementary *ex situ* conservation actions for this species such as collection and storage of seeds preserved in João do Amaral Franco Seed Bank (Ajuda Botanical Garden) have also been conducted (Caperta *et al.* 2013) with the hope that the genetic diversity contained in these collections could be representative of natural populations. Thus, to conserve *L. multiflorum*, it is preferable to develop small, localized experimental actions as Laguna *et al.* (2013) pointed out. In fact, only moderate success is reported for reintroduction projects due, for example, to insufficient long-term monitoring following reintroduction and/or lack of understanding of the underlying reasons for decline in existing plant populations (e.g. Godefroid *et al.* 2011).

CONCLUSIONS

According to Rabinowitz' scheme (1981), in which species were classified into categories according to their geographic range, habitat specificity and local population size, *L. multiflorum* can be classified as a 'classic rarity' since it presents both narrow geographic range and narrow habitat specificity, thus being considered a restricted endemic. Furthermore, species' populations observed throughout its entire range consistently present small sizes and low levels of genetic diversity as revealed by population genetic and epigenetic studies using MSAP markers (Róis *et al.* 2013). Taking the above into consideration, as well as diverse social-economic impacts exerted on its populations, we considered that *L. multiflorum* deserves a status of 'Vulnerable' in the IUCN red list.

As already pointed out for other 'taxonomically complex groups' (Ennos *et al.* 2005), conservation of rare and endemic *L. multiflorum* is best achieved by facilitating evolutionary interactions among its members that generate and maintain their taxonomic biodiversity.

SOURCES OF FUNDING

Our work was funded by Fundação para a Ciência e Tecnologia (FCT) (Portugal) (PEST-OE/AGR/UI0240/2011). A.D.C. (Researcher, CBAA/ISA) and A.S.R. (SFRH/BD/62542/2009 grant) were supported by FCT.

ACKNOWLEDGEMENTS

We thank Manuela Rodrigues (Volunteer), Ana Cortinhas (CBAA/ISA) and Vanessa Mendonça (CBAA/ISA) for help in field work, Jorge Cadima (ISA) for help in defining the experimental design, and the anonymous reviewers for all the helpful comments and suggestions. We thank the Instituto da Conservação da Natureza e das Florestas (ICNF) for providing permissions for field work.

REFERENCES

1. Akeroyd J, Wyse Jackson P. A handbook for botanic gardens for reintroduction of plants to the wild. Richmond: BGCI; 1995.
2. APA [The Environmental Protection Agency of Portugal]. Guide for the assessment of imminent threats and environmental damages. Lisboa: The Ministry of Agriculture, the Sea, The Environment and Territorial Planning; 2011.
3. Asensi A, Diez Garretas B, van der Maarel E. Dry coastal ecosystems of Portugal. In: van der Maarel E, editor. Dry coastal ecosystems: polar regions and Europe. Ecosystems of the World 2A. Amsterdam: Elsevier; 1993. p. 342-347.
4. Baumberger T, Affre L, Torre F, Vidal E, Dumas PJ, Tatoni T. Plant community changes as ecological indicator of seabird colonies' impacts on Mediterranean Islands. Ecological Indicators 2012;15:76-84.
5. Bloom WL. Origin of reciprocal translocations and their effects in *Clarkia speciosa*. Chromosoma 1974;49:61-76.
6. Böcher T. Experimental taxonomic studies in the *Arabis holboellii* and allied species. Biologiske Skrifter 1954;14:1-74.
7. Borchers-Kolb E. Ranunculus sect. Auricomus in Bayern und den angrenzenden Gebieten. I. Allgemeiner Teil. Mitteilungen der Botanischen Staatssammlung München 1983;19:363-429.
8. Braun-Blanquet J, Rozeira A, Pinto da Silva AR. Résultats de deux excursions géobotaniques à travers le Portugal septentrional et moyen IV. Esquisse sur la végétation dunale. Agronomia Lusitana 1972;33:217-234.
9. Brussard PF. The role of ecology in biological conservation. Ecological Applications 1991;1:6-12.
10. Campos JA, Herrera M, Biurrun I, Loidi J. The role of alien plants in the natural coastal vegetation in central-northern Spain. Biodiversity and Conservation2004;13:2275-2293.
11. Caperta AD, Paes AP, Espírito-Santo MD, Silva V, Ferreira A, Saraiva S. Arribas da faixa costeira de Cascais. Projecto de Recuperação de Habitats. El Botânico2013;7:69-71.

12. Castro M, Rosseló JA. Karyology of *Limonium* (Plumbaginaceae) species from the Balearic Islands and the western Iberian Peninsula. Botanical Journal of the Linnean Society 2007;155:257-272.
13. Conser C, Connor EF. Assessing the residual effects of *Carpobrotus edulis*invasion, implications for restoration. Biological Invasions 2009;11:349-358.
14. Costa JC. Tipos de Vegetação e adaptações das plantas do litoral de Portugal continental. In: Moreira ME, Moura A, Granja H, Noronha F, editors. Homenagem (In Honorium) Professor Doutor Gaspar Soares de Carvalho. Braga: Universidade do Minho; 2001. p. 283-299.
15. Costa JC, Aguiar C, Capelo JH, Lousã M, Neto C. Biogeografia de Portugal Continental. Quercetea 1998a;0:5-56.
16. Costa JC, Capelo J, Lousã M, Espírito-Santo MD. Sintaxonomia da vegetação halocasmofítica das falésias marítimas portuguesas (Crithmo-Staticetea Br.-Bl. 1947). Itinera Geobotanica 1998b;11:227-247.
17. Costa JC, Capelo J, Aguiar C, Neto C, Lousã M, Espírito-Santo MD. An overview of the Pegano harmalae-Salsoletea vermiculatae Br.-Bl. and. O. Bolòs 1958, vegetation class in continental Portugal. Colloques Phytosociologiques 2000;27:81-93.
18. Costa JC, Monteiro-Henriques T, Neto C, Arsénio P, Aguiar C. The application of the Habitats Directive in Portugal. Fitosociologia 2007;44:23-28.
19. Costa JC, Neto C, Aguiar C, Capelo J, Espírito-Santo MD, Honrado J, Pinto-GomesC, Monteiro Henriques T, Menezes de Sequeira M, Lousã M. Vascular plant communities in Portugal (Continental, the Azores and Madeira). Global Geobotany2012;2:1-180.
20. Council Directive 92/43/EEC. On the conservation of natural habitats and of wild fauna and flora. Official Journal of the European Union L 21 May 1992;206:7-50.
21. D'Antonio CM. Seed production and dispersal in the non-native, invasive succulent *Carprobrotus edulis* (Aizoaceae) in coastal strand communities of central California. Journal of Applied Ecology 1990;27:693-702.
22. D'Antonio CM, Mahall BE. Root profile and competition between

the invasive, exotic perennial, *Carprobrotus edulis* and two native species in California coastal shrubs. American Journal of Botany 1991;78:885-894.

23. Domínguez Lozano F, Moreno Saiz JC, Sainz Ollero H. Rarity and threat relationships in the conservation planning of Iberian flora. Biodiversity and Conservation 2003;12:1861-1882.
24. EC. European Commission: seventh updated list of sites of community importance for the Mediterranean biogeographical region. Official Journal of the European Union L 2013;350:101-198.
25. Ennos RA, French GC, Hollingsworth PM. Conserving taxonomic complexity.Trends in Ecology and Evolution 2005;20:164-168.
26. Erben M. Die gattung Limonium im südwestmediterranen Raum. Mitteilungen der Botanischen Staatssammlung München 1978;14:361-631.
27. Erben M. Flora Iberica, Vol. 3. Plumbaginaceae (partim)–Capparaceae. Madrid:Real Jardín Botánico, CSIC; 1993. *Limonium* Miller; p. 2-143. Castroviejo S, Aedo C, Cirujano S, Laínz M, Montserrat P, Morales R, Muñoz Garmendia F, Navarro C, Paiva J, Soriano C.
28. Espírito-Santo MD, Alves HN, Caperta AD, Moreira I. Plantas endémicas do litoral de Portugal Continental. In: Monteiro A, Gomes da Silva F, Jorge R, editors. Gestão e Conservação da Flora e da Vegetação de Portugal e da África Lusófona. In Honorium do Professor Catedrático Emérito Ilídio Rosário dos Santos Moreira. Lisboa: ISA Press; 2012. p. 267-302.
29. Farnsworth EJ. Plant life history traits of rare versus frequent plant taxa of sandplains: implications for research and management trials. Biological Conservation 2007;136:44-52.
30. Frederiksen L, Kollmann J, Vestergaard P, Bruun HH. A multivariate approach to plant community distribution in the coastal dune zonation of NW Denmark.Phytocoenologia 2006;36:321-342.
31. Godefroid S, Piazza C, Rossi G, Buord S, Stevens AD, Aguraiuja R, Cowell C, Weekley CW, Vogg G, Iriondo J, Johnson I, Dixon B, Gordon D, Magnanon S, ValentinB, Bjureke K, Koopman

R, Vicens M, Virevaire M, Vanderborght T. How successful are plant species reintroductions? Biodiversity and Conservation 2011;144:672-682.

32. Grigore M-N, Ivanescu L, Toma C. Halophytes: an integrative anatomical study.New York: Springer; 2014.

33. HeywoodVH, Iriondo JM. Plant conservation: old problems, new perspectives.Biological Conservation 2003;113:321-335.

34. Hörandl E, Greilhuber J, Klimova K, Paun O, Temsch E, Emadzade K, Hodálová I. Reticulate evolution and taxonomic concepts in the *Ranunculus auricomus* complex (Ranunculaceae): insights from morphological, karyological and molecular data.Taxon 2009;58:1194-1215.

35. ICNF. Relatório Nacional de Aplicação da Directiva Habitats (2007–2012). Lisboa:Instituto da Conservação da Natureza e das Florestas; 2013.

36. IUCN. 2013. *IUCN red list of threatened species*. Version 2013.2http://www.iucnredlist.org 5 February 2014.

37. James SH. Complex hybridity in *Isotoma petraea*. 1. The occurrence of interchange heterozygosity, autogamy and a balanced lethal system. Heredity1965;20:341-353.

38. Khan Z, Santpere G, Traveset A. Breeding system and ecological traits of the critically endangered endemic plant *Limonium barceloi* (Gil and Llorens) (Plumbaginaceae). Plant Systematics and Evolution 2012;298:1101-1110.

39. Kubitzki K. Plumbaginaceae. In: Kubitzki K, Rohwer JG, Bittrich V, editors. The families and genera of vascular plants. Vol. 2. Berlin: Springer; 1993. p. 523-530.

40. Laguna E, Deltoro V, Ferrer PP, Navarro A, Ferrando I, Escribá MC, Albert FJ. Relevant plant recovery programmes. Conservation management of plant micro-reserves and ecological restoration. In: Kadis C, Thanos CA, Laguna E, editors. Plant micro-reserves: from theory to practice. Experiences gained from EU LIFE and other related projects. Athens: Utopia Publishing; 2013. p. 127-140.

41. Lledó MD, Crespo MB, Fay MF, Chase MW. Molecular phylogenetics of *Limonium*and related genera (Plumbaginaceae): biogeographical and systematic implications.American Journal of Botany 2005;92:1189-1198.

42. Magos-Brehm J, Maxted N, Ford-Lloyd BV, Martins-Loução MA. National inventories of crop wild relatives and wild harvested plants: case-study for Portugal.Genetic Resources and Crop Evolution 2008;55:779-796.
43. Martins MC, Neto C, Costa JC. The meaning of mainland Portugal beaches and dunes' psammophilic plant communities: a contribution to tourism management and nature conservation. Journal of Coastal Conservation 2013;17:279-299.
44. Maun MA. Burial of plants as a selective force in sand dunes. In: Martínez ML,Psuty NP, editors. Coastal dunes. Ecology and conservation. Heidelberg: Springer;2004. p. 119-135.
45. Menezes de Sequeira M, Espírito-Santo MD, Aguiar C, Capelo J, Honrado J, coords. Checklist da Flora de Portugal (Continental, Açores e Madeira). Lisboa: ALFA [Associação Lusitana de Fitossociologia]; 2011.
46. Mesquita S, Sousa AJ. Bioclimatic mapping using geostatistical approaches: application to mainland Portugal. International Journal of Climatology2009;29:2156-2170.
47. Molinari HBC, Marur CJ, Daros E, Campos MKF, Carvalho JFRP, Filho JCB, PereiraLFPP, Vieira LGE
48. . Evaluation of the stress-inducible production of proline in transgenic sugarcane (*Saccharum* spp): osmotic adjustment, chlorophyll fluorescence and oxidative stress. Physiologia Plantarum 2007;130:218-229.
49. Monteiro-Henriques T. Lisboa: Universidade Técnica de Lisboa, Instituto Superior de Agronomia; 2010. *Fitossociologia e Paisagem da Bacia Hidrográfica do Rio Paiva/Landscape and phytosociology of the Paiva River's hydrographical basin*. PhD Thesis.
50. Mucina L. Europe, ecosystems of. In: Levin SA, editor. Encyclopedia of biodiversity. Waltham, MA: Academic Press; 2013. p. 333-346.
51. Palacios C, Rosselló JA, González-Candelas F. Study of the evolutionary relationships among *Limonium* species (Plumbaginaceae) using nuclear and cytoplasmic molecular markers.MolecularPhylogeneticsandEvolution2000;14:232-249.

52. Palop-Esteban M, Segarra-Moragues JG, González-Candelas F. Historical and biological determinants of genetic diversity in the highly endemic triploid sea lavender *Limonium dufourii* (Plumbaginaceae). Molecular Ecology 2007;16:3814-3827.
53. Pärtel M, Kalamees R, Reier U, Tuvi E-L, Roosaluste E, Vellak A, Zobel M. Grouping and prioritization of vascular plant species for conservation: combining natural rarity and management need. Biological Conservation 2005;123:271-278.
54. Pinto-Ricardo C, Espírito-Santo MD, Carrasquinho I, Veloso C, Ricardo CP. Impact of the invasive *Carpobrotus edulis* on *Armeria pseudarmeria* at Cabo da Roca.Proceedings of the XIII OPTIMA Meeting. 2010. p. 85.
55. Pyšek P, Genovesi P, Pergl J, Monaco A, Wild J. Plant invasions in protected areas: patterns, problems and challenges, invading nature. Vol. 7. Dordrecht; 2013. Plant invasions of protected areas in Europe: an old continent facing new problems.Foxcroft LC, Pyšek P, Richardson DM, Genovesi P Springer Series in Invasion Ecology.
56. R Core Team. R: a language and environment for statistical computing. Vienna, Austria: R Foundation for Statistical Computing; 2013. http://www.R-project.org.
57. Rabinowitz D. Seven forms of rarity. In: Synge H, editor. The biological aspects of rare plant conservation. New York: John Wiley & Sons Ltd; 1981. p. 205-217.
58. Rivas-Martínez S. Mapa de series, geoseries y geopermaseries de vegetación de España. Memoria del mapa de vegetación potencial de España. Itinera Geobotanica2007;17:5-436.
59. Rodríguez S, Palop ML, Palacios C, González-Candelas F. Molecular and morphological differentiation in *Limonium dufourii* (Plumbaginaceae), an endangered Mediterranean plant. Conservation Genetics 2003;4:383-391.
60. Róis AS, Teixeira G, Sharbel TF, Fuchs J, Martins S, Espírito-Santo MD, Caperta AD. Male fertility versus sterility, cytotype, and DNA quantitative variation in seed production in diploid and tetraploid sea lavenders (*Limonium* sp., Plumbaginaceae) reveal diversity in reproduction modes. Sexual Plant Reproduction 2012;25:305-318.

61. Róis AS, López CMR, Cortinhas A, Erben M, Espírito-Santo MD, Wilkinson M, Caperta AD. Epigenetic rather than genetic factors may explain phenotypic divergence between coastal populations of diploid and tetraploid (*Limonium* sp.; Plumbaginaceae) in Portugal. BMC Plant Biology 2013;13:205.
62. Salisbury E. Downs and dunes. Their plant life and environment. London: G. Bell and Sons; 1952.
63. Salvini D, Fineschi S, Pastorelli R, Sebastiani F, Vendramin G. Absence of geographic structure in European populations of *Rubus fruticosus* L. complex using chloroplast DNA microsatellites. Journal of the American Society for Horticultural Science 2006;131:616-621.
64. Scheffers A, Kelletat D. Tsunami relics on the coastal landscape west of Lisbon, Portugal. International Journal of the Tsunami Society 2005;1:3-16.
65. Schemske DW, Husband BC, Ruckelshaus MH, Goodwillie C, Parker I, Bishop JG. Evaluating approaches to the conservation of rare and endangered plants. Ecology1994;75:584-606.
66. Severns PM, Liston A. Intraspecific chromosome number variation: a neglected threat to the conservation of rare species. Conservation Biology 2008;22:1641-1647.
67. Sieber VK, Murray BG. Structural and numerical chromosomal polymorphism in natural populations of *Alopecurus* L. (Poaceae). Plant Systematics and Evolution1981;139:121-136.
68. Simberloff D. The contribution of population and community to conservation science. Annual Review of Ecology and Systematics 1988;19:473-511.
69. Snogerup S. Evolutionary and plant geographical aspects of chasmophytic communities. In: Davis PH, Harper PC, Hedge IC, editors. Plant life of South West Asia. Aberdeen: Botanical Society of Edinburgh UK; 1971. p. 157-170.
70. Sokal RR, Rohlf FJ. Biometry: the principles and practice of statistics in biological research. 3rd edn. New York: WH Freeman and Co; 1997.
71. Suárez-García C, de Paz JP, Febles R, Caujapé-Castells J. Genetic diversity and floral dimorphism in *Limonium dendroides*

(Plumbaginaceae), a woody *Canarian*species on the way of extinction. Plant Systematics and Evolution 2009;280:105-117.

72. Suehs CM, Médail F, Affre L. Invasion by South African *Carpobrotus* (Aizoaceae) taxa in the Mediterranean Basin: the effects of insularity on plant reproductive systems. In: Child L, Brock JH, editors. Plant invasions: ecological threats and management solutions. Leiden: Backhuys Publishers; 2003. p. 247-263.
73. Traveset A, Brundu G, Carta L, Mprezetou I, Lambdon P, Manca M, Médail F, Moragues E, Rodríguez-Pérez J, Siamantziouras AD, Suehs CM, Troumbis AY, Vilà M,Hulme PE. Consistent performance of invasive species within and among islands of the Mediterranean basin. Biological Invasions 2008;10:847-858.
74. van der Maarel E. Dry coastal ecosystems: polar regions and Europe. The Netherlands: Elsevier Science Publishers; 1993.
75. van der Maarel E, van der Maarel-Versluys M. Distribution and conservation status of littoral vascular plant species along the European coasts. Journal of Coastal Conservation 1996;2:73-92.
76. van Dijk PJ. Ecological and evolutionary opportunities of apomixis: insights from Taraxacum and Chondrilla. Philosophical Transactions of the Royal Society of London. Series B, Biological Sciences 2003;358:1113-1121.
77. Vilà M, Tessier M, Suehs CM, Brundu G, Carta L, Galanidis A, Lambdon P, MancaM, Médail F, Moragues E, Traveset A, Troumbis AY, Hulme PE. Local and regional assessment of the impacts of plant invaders on vegetation structure and soil properties of Mediterranean islands. Journal of Biogeography 2006;33:853-861.
78. Ward JH. Hierarchical grouping to optimize an objective function. Journal of the American Statistical Association 1963;58:236-244.
79. Warming E. Oecology of plants: an introduction to the study of plant-communities. Oxford: Clarendon Press; 1909.
80. Young AG, Murray BG. Genetic bottlenecks and dysgenic gene flow into re-established populations of the grassland daisy, *Rutidosis leptorrhynchoides*. Australian Journal of Botany 2000;48:409-416.
81. Zar JH. Biostatistical analysis. Prentice-Hall: Upper Saddle River; 2010.

Chapter 5

Floral Specialization and Angiosperm Diversity: Phenotypic Divergence, Fitness Trade-Offs and Realized Pollination Accuracy

W. Scott Armbruster[1,2,3,*]

[1]School of Biological Sciences, University of Portsmouth, Portsmouth PO1 2DY, UK

[2]Institute of Arctic Biology, University of Alaska Fairbanks, Fairbanks, AK 99775-7000, USA

[3]Department of Biology, Norwegian University of Science & Technology, Trondheim N-7491, Norway

ABSTRACT

Plant reproduction by means of flowers has long been thought to promote the success and diversification of angiosperms. It remains unclear, however, how this success has come about. Do flowers, and their capacity to have specialized functions, increase speciation rates or decrease extinction rates? Is floral specialization fundamental or incidental to the diversification? Some studies suggest that the conclusions we draw about the role of flowers in the diversification and increased phenotypic disparity (phenotypic diversity) of angiosperms depends on the system. For orchids, for example, specialized pollination may have increased speciation rates, in part because in most orchids pollen is packed in discrete units so that pollination is precise enough to contribute to reproductive isolation. In most plants, however, granular pollen results in low realized pollination precision, and thus key innovations involving flowers more likely reflect reduced extinction rates combined with opportunities for evolution of greater phenotypic disparity (phenotypic diversity) and occupation of new niches. Understanding the causes and consequences of the evolution of specialized flowers requires knowledge of both the selective regimes and the potential fitness trade-offs in using more than one pollinator functional group. The study of floral function and flowering-plant diversification remains a vibrant evolutionary field. Application of new methods, from measuring natural selection to estimating speciation rates, holds much promise for improving our understanding of the relationship between floral specialization and evolutionary success.

INTRODUCTION

Flowering plants are the most abundant and diverse autotrophic organisms on land. Reproduction by means of flowers and fruits is often invoked as one of the main causes of this evolutionary success (Stebbins 1974; Regal 1977), but the mechanisms of the putative causal link between various angiosperm innovations and the group's evolutionary success remain elusive (Crepet and Niklas 2009). By evaluating both old and new ideas and identifying possible misconceptions, the following review attempts to come to grips with the role of pollinators, flowers and their interactions in the diversification and evolutionary success of plants. The evolutionary success and ecological dominance of angiosperms have been associated with a number of features. Perhaps most often invoked is their use of animals to transport pollen between flowers. The capacity of extreme specialization in flower form and function is thought to have increased diversification rates (Grant 1949; Crepet and Niklas 2009) and increased the range of pollen and seed dispersal, opening new opportunities for further specialization and diversification (Stebbins 1974). These venerable ideas need further scrutiny in order to develop a programme for testing the validity of their components.

In more recent years, discussions about the types, causes and effects of floral specialization and pollination syndromes have come to the fore. At the core of many of these arguments lies uncertainty about the fundamental concept of plant fitness trade-offs, where adaptation to one type of pollinator incurs costs in terms of reduced effectiveness of another type (Aigner 2001, 2004, 2006). Is this ubiquitous, common or rare? Consensus, if there is any, seems to be that trade-offs are frequently weak or absent (see Castellanos *et al.* 2004; Muchhala 2007), although there are probably too few studies for conclusions to be drawn.

Below I address these issues by discussing pollination and floral specialization in relation to fitness trade-offs. I also review adaptive accuracy and floral precision as they relate to mechanisms of reproductive isolation, diversification and evolution of phenotypic disparity (morphological diversity).

Evolutionary Success: Diversity and Phenotypic Disparity

Evolutionary success is commonly measured as the number of species (richness or diversity) in a clade, sometimes in combination with the ecological/morphological variation ('phenotypic disparity') of those species (Hunter 1998; Schluter 2000). [Phenotypic disparity is also called 'morphological' or 'phenotypic diversity', but I use 'disparity' here to follow a tradition already well established in the zoological literature (e.g.Wills *et al.* 1994; Foote 1997; Harmon *et al.* 2003; Hughes *et al.* 2013) and to keep a clear distinction from species diversity.] Lineages with many extant species are considered more successful than ones with very few. Lineages exhibiting greater phenotypic disparity (phenotypic diversity) are expected to be more successful because they may have greater capacity for further diversification (although the data supporting this expectation are few). The combination of relatively rapid diversification with greatly increased disparity (phenotypic diversity) is often referred to as 'adaptive radiation' (Simpson 1953; Schluter 2000). In turn, we can assess traits (character states) associated with evolutionary success, which are called 'key innovations', e.g. powered flight in insects, birds and bats, or, perhaps, flowers in flowering plants (e.g. Simpson 1944, 1953; Hodges and Arnold 1995; Hunter 1998).

There are three distinct processes that can lead to differences in evolutionary success in association with key innovations (phenotypes), the first of which is microevolutionary and the other two are essentially macroevolutionary: (i) differential trait transition rates, (ii) differential extinction rates, and (iii) differential speciation rates (Fig. 1). To illustrate the operation of these processes in more detail, consider a clade of plants where three-quarters of the species have flowers with petals, even though the basal condition is for flowers to lack petals. If petals are adaptive (say attracting more pollinators, increasing reproductive success), there may be differential transitions between states (Fig. 1), such that lineages with flowers lacking petals sometimes evolve them, but once evolved, they are rarely lost (i.e. lineages with petaliferous flowers almost never revert to being apetalous). Thus petals are a key innovation in an adaptive (microevolutionary) sense: they enhance reproductive success of individuals and populations, which in turn causes frequent microevolutionary transitions to the state. If we have enough phylogenetic information, we may recognize the pattern this process creates: parallelism (i.e. multiple origins of petals). However, multiple transitions to having petals could also be cryptic if the taxa are closely related, the phylogenetic tree is poorly resolved, taxa are missing or intermediate lineages have gone extinct. There might be only one origin detected, when actually several transitions have occurred.

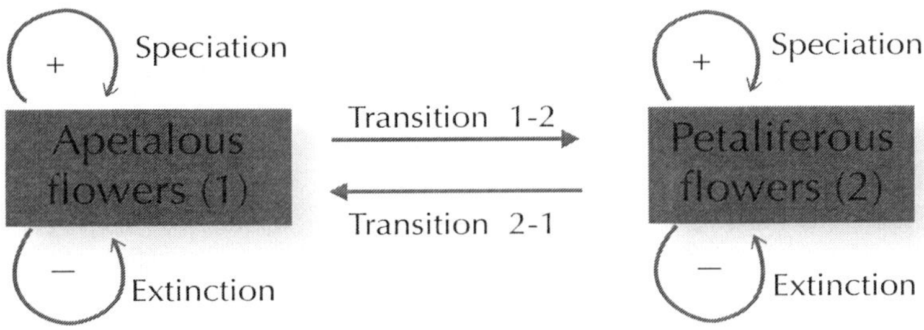

Figure 1: Causes of differences in species diversity of lineages with different character states, here, for example, bearing flowers with (petaliferous) and without (apetalous) petals.

Differences in the number of petaliferous and apetalous species in a clade can also come about through differential extinction. Strictly speaking this is a macroevolutionary property, for example, as when the biological 'decks are cleared' by astronomically induced mass extinctions (e.g. by large-meteor impacts). However, if extinction has occurred over extended periods (sometimes called the 'background extinction' rate), not in a single pulse, there is a likely link to normal microevolutionary processes. Thus if petals attract more pollinators, resulting in better pollination and lower rates of reproductive failure, lineages with petals would tend to experience lower probabilities of extinction.

Finally, if petals promote attraction of particular kinds of pollinators, the presence of petals might increase speciation rates by promoting pre-pollination reproductive isolation from related taxa. This could come about in three ways: (i) attracting and rewarding a set of pollinator species not used by related plant species, (ii) petals enforcing specialized handling, promoting associative learning or otherwise enhancing constancy (temporary floral specialization by individual animals), or (iii) petals causing pollen placement and stigma contact in places on the pollinator not used by other species. Verne Grant (1949, 1971, 1994*a*) classified (i) and (ii) as ethological isolation and (iii) as mechanical isolation.

Semantic Issues

Pollinator Importance, Effectiveness, Efficiency and Abundance

Stebbins (1970, 1974) proposed the 'most effective pollinator principle' which states that:Since selection is a quantitative process, the characteristics of the flower will be molded by those pollinators that visit it most frequently and effectively in the region where it is evolving.

Although the intention of this statement is clear in this context, Stebbins' unfortunate use of 'effective' both in the name of the principle and as a component of importance in natural selection has led to confusion about which pollinators are expected to be driving selection. My interpretation of what Stebbins meant in this statement can be expressed by changing one word:'... the characteristics of the flower will be molded by those pollinators that visit it most frequently and [*efficiently*] ...'

(i.e. the most *effective pollinators*). [As noted below, we should probably also insert 'usually' in this statement; when fitness trade-offs in pollinator use are absent, the 'moulding' pollinator may not be the most effective one (Aigner 2001).]

Thus I recommend we define pollinator effectiveness as the product (or a similar function) of visitation rate (frequency) and per-visit efficiency (measured as pollen delivered, seeds produced, offspring sired, etc., per visit; see also Freitas 2013). For example,

$$\text{Pollinator effectiveness} = \text{visitation rate} \times \text{efficiency} \qquad (1)$$

Earlier my colleagues and I have used 'pollinator importance' as a less ambiguous term, but conceptually related to effectiveness as defined above (Armbruster 1985, 1988, 1990; Armbruster *et al.* 2000; Fenster *et al.* 2004). We referred to the pollinator with the highest importance score as the 'principal pollinator'. One practical estimator of pollinator importance (PI) is

$$PI = V \times A \times S \qquad (2)$$

where V is the visitation rate per unit time, A is the per-visit probability of contacting the anthers and S is the per-visit probability of contacting the stigmas (Armbruster 1985, 1988, 1990). Freitas (2013) has come to a similar conclusion about the confusion caused by these semantics. He recommends calling visitation rate 'visitation intensity' and efficiency 'efficacy', but otherwise his recommendations largely agree with the above.

Floral Isolation

Before evaluating the role of floral specialization in diversification, it is necessary to define floral isolation precisely. Considerable confusion exists in the literature because the term has been used in at least two ways (see Fig. 2): (i) floral isolation *sensu stricto* (*s.s.*), which implies biologically significant reproductive isolation as a result of floral traits acting on their own (i.e. floral isolation as a category of reproductive isolation; see Grant 1949, 1971, 1994*a*, *b*; reviewed in Kay and Sargent 2009); (ii) floral 'isolation' *sensu lato* (*s.l.*), where there is some segregation of pollen flow, i.e. more intraspecific and less interspecific, without necessarily resulting in biologically significant reproductive isolation (e.g. Huang and Shi 2013; see the discussion in Kay and Sargent 2009). One problem with this dichotomy is: what is meant by 'biologically significant', and can it ever be defined or assessed? For the purpose of this review, I will treat 'biologically significant isolation' to be isolation that generates segregation of gene flow sufficient to allow, by itself, genetic divergence of the population in response to drift or weak selection. Thus narrow-sense floral isolation acting by itself generates sufficient reproductive isolation to prevent or retard genetic homogenization of populations. In contrast, broad-sense floral 'isolation' includes anything that reduces inter-morph mating even by a few percentage points, and may or may not have any direct genetic consequences. Because I think of floral isolation as a type of reproductive isolation and because reproductive isolation is commonly defined as 'the inability of a species to breed successfully with related species' (Merriam-Webster 2013), I prefer to use 'floral isolation' in the narrow sense. In place of 'floral isolation*s.l.*', I will usually refer directly to the selective mechanisms involved in reducing interspecific pollination (e.g. divergent adaptations to different pollinators, reinforcement of reproductive isolation or character displacement). Although I do not recommend abandoning the term 'floral isolation', I urge that we always explain in what sense we are using it.

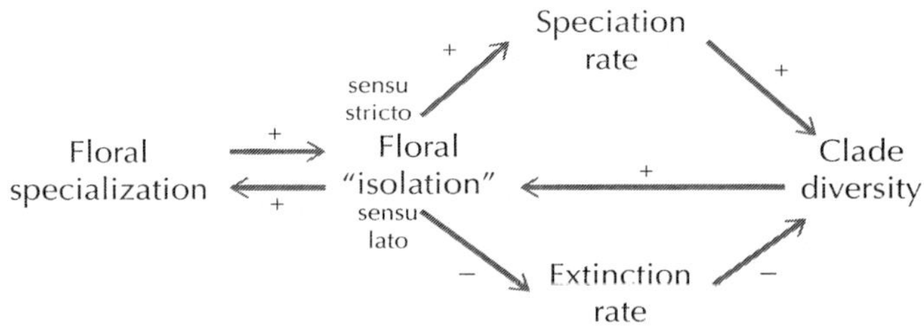

Figure 2: Complex links between floral specialization, floral 'isolation' and clade species diversity/richness. Specialized flowers may increase the likelihood of floral isolation in the strict sense and/or in the broad sense. Increased occurrence of floral isolation, in the strict sense, may increase the speciation rate and hence clade species richness and diversity. Alternatively, floral 'isolation', in the broad sense, may decrease the rate of extinction and lead to increases in clade species richness and thereby diversity. In turn, increased species diversity of clades may increase the number of clade members occurring in sympatry, thereby selecting for improvements in floral isolation in the broad sense, which may be manifested by increased floral specialization.

FLORAL SPECIALIZATION, FITNESS TRADE-OFFS AND ADAPTIVE ACCURACY

The concept of specialization has several components and meanings as applied to flowers. The manifold nature of this concept has created considerable confusion and discussion (cf. Ollerton 1996; Waser *et al.*1996; Johnson and Steiner 2000; Fenster *et al.* 2004; Ollerton *et al.* 2007).Armbruster *et al.* (2000; see also Fenster *et al.* 2004) pointed out that specialization can refer either to a state (being specialized: 'ecological specialization') or to a process (becoming more specialized: 'evolutionary specialization'). Thus, a species may be legitimately viewed as unspecialized ecologically, but specialized evolutionarily, if it is more specialized than its ancestors.

Fundamental to the definition of floral specialization is the question of what is being specialized on. Classically, specialized pollination has referred to the number of species of animals that pollinate the flowers, hence a continuous spectrum, from one species (specialized) to many species (generalized). More recently, the concept of specialization vs. generalization in pollination has focused on functional groups of pollinators. This term was apparently coined independently (but with similar meaning) by Corbet (1997—'function group'), Johnson and Steiner (2000—'function type') and Fenster *et al.* (2004—'functional group'). Pollinator functional groups are animals that generate similar selection pressures on flowers, e.g. bees with similar tongue lengths, even though they may belong to different genera or families (Johnson and Steiner 2000;Fenster *et al.* 2004). [This is similar but not the same as what Ollerton *et al.* (2007) refer to by 'functional specialization'. 'Functional specialization/generalization' *sensu* Ollerton *et al.* (2007) refers instead to the number of higher taxa being pollinators (e.g. bees, birds, bats). This contrasts with what these same authors called 'ecological specialization/generalization': the number of *species* of pollinators. 'Ecological specialization' *sensu* Ollerton *et al.* (2007) differs from the use by Armbruster *et al.* (2000), Fenster *et al.* (2004) and the use adopted here.]

Ollerton *et al.* (2007) coined the term 'phenotypic specialization' of flowers, where the morphology, colour and/or fragrance of a flower is specialized, even though the current pollinators may not be especially restricted. Thus a phenotypically specialized flower may have ecologically generalized pollination (by multiple functional groups). This unexpected outcome may reflect a holdover from a prior history of specialization, compensatory adaptations for dealing successfully with a variety of pollinators, as may be the case for *Stylidium* spp. (Armbruster *et al.* 1994,2004, 2009*a*, *b*; see below), or the lack of fitness trade-offs in adapting to pollinators that increase marginal fitness (Aigner 2001, 2004, 2006). Fitness trade-offs in this context are when a trait's positive effect on pollination effectiveness of one pollinator (fitness) creates a negative effect on pollination effectiveness of one or more other pollinators. Aigner (2001, 2006) explored the effect of fitness trade-offs in the evolution of phenotypic specialization, pointing out that, in the absence of steep fitness trade-offs, phenotypic specialization can evolve without restricting the number of pollinator types (ecological generalization is maintained; see the discussion in the next section). In contrast, in the presence of steep trade-offs, phenotypic specialization in response to selection by one pollinator species results in fewer functional types of pollinators that can be utilized (ecological and evolutionary specialization). Muchhala *et al.*(2010) explored the role of pollen fates in the evolution of floral specialization. Individual-based models indicated that male fitness differentials can drive the evolution of ecological and phenotypic specialization even in the absence of fitness trade-offs.

Fitness Trade-offs

The above studies underscore the importance of understanding how often fitness trade-offs between floral adaptations to different pollinators occur, not least because floral specialization is promoted by the fitness trade-offs between pollinators (e.g. Schemske and Horvitz 1984; Wilson and Thomson 1996; Aigner 2001; Mayfield *et al.* 2001; cf. Muchhala *et al.*2010). Aigner's own studies (2001, 2004, 2006) suggested that fitness trade-offs were absent in pollination of generalist *Dudleya* flowers by large bees, small bees and hummingbirds. Aigner (2006) reviewed a number of studies that suggested trade-offs in floral adaptation to pollination by different agents, but most of these studies either failed to elucidate the functional basis of fitness trade-offs or failed to account for both components of pollinator effectiveness (abundance and efficiency), which leaves uncertainty about the importance of the trade-offs in those systems.

An elegant experimental study with *Penstemon* (Plantaginaceae) showed no detectable fitness trade-offs across traits affecting hummingbird vs. bee pollination (Castellanos *et al.* 2004). Another experimental study, which did detect clear trade-offs, is Muchhala's (2007) study of hummingbird and bat 'pollination'. This involved artificial flowers of different widths, capturing the morphology of two species of *Burmeistera*(Campanulaceae), one primarily hummingbird pollinated (narrow floral tube) and the other primarily bat pollinated (wide floral tube). Hummingbirds were better at transferring pollen between narrow-tube 'flowers', and bats better at transferring pollen between wide-tubed 'flowers'. Importantly, intermediate-width tubes (generalists) performed worse than the narrow tubes with hummingbirds and worse than the broad tubes with bats (and had lowest 'pollination' overall), demonstrating a clear trade-off favouring two specialized phenotypes over one intermediate generalist.

Temeles *et al.* (2009) described an intriguing system involving sexually dimorphic purple-throated carib hummingbirds (*Eulampis jugularis*) exerting selection on the shape of *Heliconia* (Heliconiaceae) flowers. Using artificial flowers, they discovered trade-offs in handling time, where females, which have long, curved bills, handled longer flowers (of all curvatures) more quickly and effectively, obtaining the deeper nectar, than the males, which have short, straight bills. In turn, males had shorter handling times than females on artificial flowers with short, straight corollas (but only when hovering). Although the measured trade-offs were experienced by the pollinator, they suggested an indirect plant fitness trade-off, where both birds and plants should specialize on the appropriate morphs, with selection against generalists (Temeles *et al.*2013).

Schemske and Bradshaw (1999) and Bradshaw and Schemske (2003) found that variation in flower colour in hybrid *Mimulus* created a trade-off, affecting visitation rates of bees and hummingbirds in opposite directions. However, variation in nectar volume and petal surface area did not create trade-offs, having significant effects on only one of the two pollinator types (Schemske and Bradshaw 1999).

There is also an older literature addressing fitness trade-offs in pollination using natural variation among populations and species. Trade-offs, if any, are displayed in the shapes of estimated adaptive surfaces (see reviews inSchluter 2000; Svensson and Calsbeek 2012). Because this approach provides uncontrolled comparative results rather than controlled experimental ones, it is necessary to be more cautious in interpretation, being alert to possible confounding factors. Nevertheless, clear fitness trade-offs in *Dalechampia* (Euphorbiaceae) pollination were documented using intra-population, inter-population and inter-species comparisons to estimate the shape of the adaptive surface. These plants have bilaterally symmetrical blossoms (pseudanthial inflorescences) with pollination by resin-collecting bees (which use the resin for nest construction) of a range of sizes. Analyses showed that adaptation to pollination by large resin-collecting bees (Apidae: *Eulaema, Eufriesea*) reduced or precluded, respectively, pollination by medium-sized resin-collecting bees (Apidae:*Euglossa*) and small resin-collecting bees (Megachilidae: *Hypanthidium*;Armbruster 1985, 1988, 1990, 2006; Hansen *et al.* 2000). This trade-off operated as an interaction between (i) attraction and visitation rate, as determined by the amount of resin reward, and (2) pollinator efficiency, measured as the product of the rates of bee contact with stigmas and with anthers. This interaction creates a positive-diagonal, adaptive ridge (in 3-trait space), which is also an axis of specialization (attraction of and pollination by small bees, medium-sized bees or large bees). Off-diagonal trait combinations have either more generalized pollination with lower fitness (attraction and pollination by all bee sizes with wasteful reward investment and loss of pollen to other sympatric *Dalechampia* species) or extremely low pollination rates (attraction of only small bees when stigmas are contacted only by large bees; Fig. 3; Armbruster 1988, 1990, 2006). All populations and species were observed to occupy the predicted adaptive ridge. This ridge is an allometric trajectory, so it is possible that the relationship is pleiotropic rather than created by a fitness trade-off, although phenotypic (Armbruster 1991) and quantitative-genetic (Hansen*et al.* 2003*a*, *b*) data suggest that this is very unlikely.

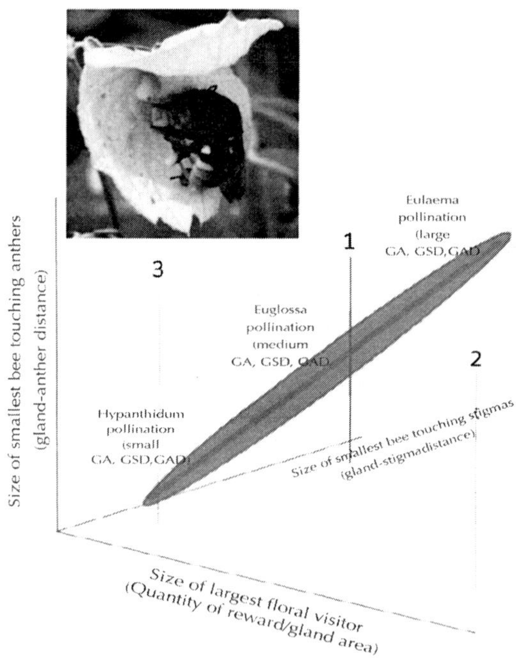

Figure 3: Adaptive 'ridge' (a series of concentric cigar-shaped volumes) in three-trait space, capturing fitness trade-offs in *Dalechampia*spp. with resin

rewards. Adaptation to small-bee pollinators precludes visitation by large bees, while adaptation to large-bee pollinators precludes small-bee pollinators. The darker the shade of blue, the higher the fitness in that volume. Off-diagonal volumes (clear) experience lowest fitness. Region 1 is a volume of low fitness because only small bees are attracted, but only large bees contact stigmas and anthers. Region 2 is a volume of low fitness because resin costs exceed pollination benefits, and/or greater interspecific pollination occurs. Region 3 is a volume of low fitness because, although the small bees attracted touch the stigmas, they do not carry pollen because they do not contact the anthers.

Another comparative study system exhibiting bilaterally symmetrical flowers and apparent fitness trade-offs is the genus *Collinsia*(Plantaginaceae). These annuals have flowers resembling pea flowers in having stamens and style enclosed in a keel-like fold of the lower corolla lobe (see Kalisz *et al.* 1999). The keel is depressed when alighting bees are of sufficient mass (in passive depression) or strength (in active depression). With depression of the keel, the stigma contacts, and/or pollen is deposited on, the underside of the bee (Fig. 4; Armbruster *et al.*2002, 2004, 2009*a*). Variation in flower size and peduncle (floral stem) strength creates continuous variation in 'optimal' pollinator size. Large-flowered species attract large bees (Apidae: *Bombus, Anthophora,* etc.) and have strong peduncles to support the weight. These flowers are also visited by smaller bees (Megachilidae: *Osmia* spp., especially males; various Halictidae), but these bees are too small to depress the keel, and they obtain nectar (nectar thieves) or glean stray pollen without contacting anthers or stigmas. Small-flowered species of *Collinsia,* in contrast, do not usually attract large bees, both because of less nectar (Heinrich and Raven 1972) and because if large bees visit small-flowered species the peduncles collapse under their weight and the bees are dumped onto the ground (an experience the bees seem to avoid). The corolla of small flowers is lightly spring-loaded so that small bees can depress the keel and act as good pollinators. The allometry of floral size and accompanying 'engineering' creates an apparent trade-off, where large flowers can be pollinated by large bees but not small ones, and small flowers can be pollinated by small bees but are seldom visited by large ones. In the absence of an experiment, one cannot rule out that the apparent trade-off could be eliminated by selection for generalized pollination, but it seems unlikely.

Figure 4: Flower of *Collinsia sparsiflora* with pollinating male *Eucera* (*Tetralonia*) bee (Apidae: Anthophorinae). The keel of the lower lip is partially depressed, with the fertile part contacting the underside of the thorax.

Perusal of published studies suggests the hypothesis that trade-offs seem more likely to occur in bilaterally symmetrical (zygomorphic) flowers than radially symmetric flowers. However, a notable exception to this trend is found in the plant genus *Stylidium* ('triggerplants', Stylidiaceae). These Australasian herbs and shrubs have phenotypically highly specialized, protandrous flowers that are zygomorphic and bear nectar in a tube or spur. The column (gynostemium) is formed by fused pistillate and staminate tissues and is motile, 'explosively' depositing pollen on, or retrieving it from, the pollinator in 10–15 ms (Fig. 5; Findlay 1978;Armbruster *et al.* 1994, 2004, 2009*a*). The flowers are ecological generalists, attracting a large number of species in several function groups of pollinators (small bees, large bees, syrphid flies, and small-, medium- and large-bodied bee flies). All functional groups are similarly effective pollinators across a range of flower sizes. The fitness contributions of pollinators are additive; there are no detectable trade-offs over the range of floral visitors in these functional groups.

This is because, except for the extremes, the flowers do not experience morphological, fit-related trade-offs, where the size of the flower must match exactly the size and shape of the pollinator for pollen to be picked up and deposited, as described above. Instead the motile column behaviourally 'adapts' to size and shape of the diverse pollinators. Pollen is deposited and stigmas make contact in the same location on a given pollinator species. However, anther/stigma contact is in different locations on different pollinator species, but consistently so: e.g. *Stylidium lineatum* places and retrieves pollen at the tip of the abdomen of medium-sized bee flies, the top of the thorax of large bee flies and the front of the head of large anthophorid bees (W. S. Armbruster, unpubl. res.). Thus the complex, specialized *Stylidium* floral morphology and trigger behaviour work equally well with a variety of sizes and shapes of pollinators. In fact the flower could be described as phenotypically specialized to be an excellent ecological generalist.

Figure 5: *Stylidium dichotomum*flower in female phase retrieving pollen from*Leioproctus* sp. which has sprung the column while obtaining nectar.

There are good reasons also to think that most open, actinomorphic (radial) flowers (e.g. *Dudleya*) and capitulae (e.g. most Asteraceae) usually lack fitness trade-offs across a broad range of pollinator functional groups. However, this is not always the case. There are apparent trade-offs in rates of contact with stigmas for actinomorphic flowers with 'apertures' of some sort, where the aperture interacts with pollinator size. For example, in some *Linum* (Linaceae) and *Parnassia*(Celastraceae/Parnassiaceae) species and in most *Passiflora*(Passifloraceae), flies and bees that are too large to fit through the gap formed by petals and fertile parts are excluded from reaching nectar and are rarely good pollinators. Excessively small flies and bees have access to the nectar but usually slip past the anthers and stigmas without making contact, thus failing to pick up or deposit pollen (high visitation, low pollination efficiency; Armbruster *et al.* 2006, 2009*a*, 2014*a*; Benevides *et al.* 2013).

Ideally, all of the above comparative systems should be tested experimentally using factorial analyses with different species of pollinators and different floral phenotypes (Wilson and Thomson 1996; Muchhala 2007). However, I think it is safe to accept, at least provisionally, the results of these comparative analyses. The existence of fitness trade-offs in certain kinds of flowers, e.g. most zygomorphic, and not others, e.g. many actinomorphic, could drive evolution towards ecological specialization and perhaps greater diversification in certain groups (e.g.Sargent 2004). Indeed, phylogenetic-comparative analyses of the evolution and diversification consequences of fitness trade-offs across pollinator types should be a fruitful line of investigation.

Conflicting Selection Generated By Abiotic Factors and Multi-Species Interactions with Flowers

It is now well recognized that plants interact with a diversity of mutualists and antagonists, from pollinators and defending ants to nectar robbers, florivores, folivores and pathogens, in both ecological and evolutionary time (Strauss and Irwin 2004; McCall and Irwin 2006; Strauss and Whittall 2006; Irwin *et al.* 2010). For example, while advertisement traits increase the apparency of flowers to pollinators, raising fitness, they may also increase the apparency to enemies such as nectar robbers, florivores and seed predators, lowering fitness. Thus floral traits may commonly be subject to conflicting selection mediated by mutualists and antagonists, resulting in complex trade-offs. Depending on the shape of the fitness responses to the conflicting relationships, the net result may often be stabilizing selection. The role of multiple species in generating stabilizing selection may help in explaining why phenotypic selection studies that address only one interaction, as is often the case, more often document directional selection than the otherwise expected stabilizing selection.

Numerous examples of conflicting selection on floral traits have been published, and several papers review the topic thoroughly (e.g. Brody 1997; Cariveau *et al.* 2004; Strauss and Irwin 2004; McCall and Irwin 2006; Strauss and Whittall 2006; Irwin *et al.* 2010). One of the earliest well-documented examples is the observation of conflicting selection on floral fragrance and flower shape of *Polemonium viscosum* generated by ants and bumble bees in the subalpine Colorado (Galen 1999; Galen *et al.*1987, 2011). A very recent example is the detection of conflicting selection on the size of floral bracts of *Dalechampia scandens* in Mexico, with positive selection generated by pollinating bees and negative selection generated by curculionid seed predators (Pérez-Barrales *et al.*2013).

Additional complexity in floral evolution may result from conflicting selection generated by pollinators and abiotic factors. For example, pollinators often select for larger corolla size, but such increases exert a large cost in terms of water loss in dry environments (Galen 2000;Lambrecht and Dawson 2007). Selection to speed up the life cycle in highly seasonal environments may result in selection against large flowers, in conflict with selection generated by pollinators (Runions and Geber 2000; Mazer *et al.* 2004; Lambrecht 2013; but see also Moeller and Geber 2005).

Adaptive Accuracy of Flowers

Another approach to assessing fitness trade-offs and possible connections between floral specialization and flowering-plant diversification involves adaptive accuracy theory.

This approach, which can be used in this context to quantify pollination accuracy, is also useful for assessing the ability of the flowers of sympatric species to generate segregated pollen flow and maintain reproductive isolation (or not).

Adaptive inaccuracy estimates the phenotypic load (maladaptation) that results from phenotypic departure from the optimum in a population. At the level of the population, there are at least three components, which are additive (Armbruster *et al.* 2004, 2009*a*; Hansen *et al.* 2006; Pélabon *et al.*2012): (i) optimality of the mean, which is how far the mean of events departs from the optimum (='maladaptive bias'), (ii) the variance, which is how much individuals vary from the mean (='adaptive imprecision'), and (iii) the variance in the optimum. By extrapolation from measurement theory (Armbruster *et al.* 2004; Pélabon *et al.* 2012), these three components sum to the adaptive inaccuracy as

Adaptive inaccuracy=(trait mean – optimum value)^2+trait variance + optimum variance (3)

Although phenotypic selection only detects relative overall adaptive inaccuracy of flowers (as a component of reproductive fitness; Armbrusteret *al.* 2004, 2009*a*, *b*), genetic response to phenotypic selection may occur through reducing the maladaptive bias (less departure from the optimum), decreasing (or increasing) the adaptive imprecision, decreasing (or increasing) the variance in the optimum or some combination of all three.

There are two new aspects of adaptive accuracy that need to be considered in the context of pollination: (i) *fundamental* accuracy and precision, and (ii) *realized* accuracy and precision (terms were derived by analogy to ecological niche concepts; see also Armbruster *et al.* 2014*b*).*Fundamental pollination accuracy* relates to measurements of optimality, precision and accuracy taken from the flower itself. This is only a predictor of the actual accuracy in play ecologically, the *realized pollination accuracy*. The latter then refers to the interaction of floral parts with the pollinator itself. It reflects the effects of the pollinators' behaviour in interacting with the flower (e.g. variation in approach and landing), the distribution and redistribution of pollen on the pollinator and effects of other ecological factors (e.g. other pollinators, florivores, predators). The realized precision of pollination is nearly always lower than the fundamental precision, and therefore this asymmetry also holds for accuracies.

MECHANISMS OF DIVERSIFICATION

From the overview presented in the Introduction, it is clear that flowers may influence diversification in several ways, involving both adaptive (trait transitions, reduced extinction rates, adaptive/ ecological speciation) and non-adaptive processes (e.g. non-adaptive speciation). Because the influence of adaptive processes on 'ecological' and 'adaptive' speciation has been reviewed recently elsewhere (Dieckmann *et al.* 2004; Johnson 2006, 2010; Nosil 2012; van der Niet and Johnson 2012), it is not covered in detail here. However, it should be remembered that the effect on plant speciation of adaptation to pollinators is potentially of great importance (Johnson 2006, 2010; van der Niet and Johnson 2012). In both allopatry and sympatry, adaptive divergence of floral traits can lead indirectly to the establishment of barriers to interbreeding. Additionally, adaptive reinforcement of reproductive isolation ('Wallace effect'; Grant 1966;Silvertown *et al.* 2005) and reproductive character displacement (Grant 1972) may further promote speciation rates and hence diversification. In contrast, non-adaptive speciation, even if rare, is of conceptual importance because it potentially decouples species-level selection (Stanley 1975) from microevolution (Gould 2002). The commonest form of non-adaptive speciation (in the sense of the biological-species definition) in plants is probably polyploidy. Non-adaptive ('instantaneous') speciation has also been suggested for the origins of fragrance 'races' and new species in male-euglossine-pollinated orchids and sexually deceptive orchids (Dressler 1968; Cozzolino and Widmer 2005).

Specialization and Clade Success

Features that promote specialization are commonly interpreted as key innovations, although there are reasons to be cautious in making this assumption. For example, dependence on a single obligate mutualist may greatly increase the risk of extinction (Waser *et al.* 1996). Nevertheless, the relationship between specialization and clade 'success' has fascinated biologists since Darwin. Botanists and zoologists have often thought about the relationship between these two properties somewhat differently. Zoologists have largely considered specialization to be adaptive, leading to adaptive evolution and evolutionary success (higher transition rates and lower extinction rates; e.g. Schmidt-Kittler 2002; Fernández-Marín *et al.*2004; Litman *et al.* 2011; Eastman *et al.* 2013). Surprisingly, evolutionary botanists, in contrast, have probably written more, at least implicitly in early literature, about non-adaptive processes such as non-adaptive speciation, where, for example, use of different pollinators is thought to lead incidentally to reproductive isolation and thereby increase speciation rates (Grant 1949, 1971, 1994*a*; Hodges and Arnold 1995; Hodges 1997;Sargent 2004; but see Stebbins 1974; Johnson 2006). (Ecological speciation is included in this argument, but it is less clear how adaptive specialization on different pollinators affects speciation rates.)

The role of floral specialization in reducing extinction rates is rarely discussed, although Hodges and Arnold (1995) mention 'increased reproductive success' as a possible outcome of floral specialization. Research on the role of plant physiology and leaf and stem anatomy and function in evolutionary success also emphasizes the adaptive nature of specialization (and implicitly reduced extinction; Gianoli 2004; Klak *et al.*2004; Bakker *et al.* 2005; Donoghue 2005; Guzman *et al.* 2009; Givnish 2010). Although Johnson and colleagues (Johnson 2006, 2010; van der Niet *et al.* 2006; van der Niet and Johnson 2012) have emphasized the role of adaptation to pollinators in both allopatric and sympatric divergence, it is largely in the context of speciation rather than reproductive success or population viability.

It seems clear that floral specialization is often associated with clade success (species richness; Hodges and Arnold 1995; Hodges 1997;Sargent 2004; Kay and Sargent 2009). However, it is less clear what causes this association (see below; Armbruster and Muchhala 2009).

Reduced Extinction

As already noted, it is commonly suggested that flowering plants are successful and species rich because reproduction by means of flowers and fruits is more effective overall than by means of strobili (e.g. Stebbins 1974). Indeed, typical angiosperm flowers provide numerous potential advantages over gymnosperm strobili. These include more rapid development, which may have been important in colonizing and reproducing in comparatively short-lived or highly seasonal habitats. Flowers are usually smaller than ovulate strobili, which allows greater flexibility in reproductive investment. Additionally, most flowers are hermaphroditic, leading to efficiencies in pollination by animals, as well as potential reproductive assurance by autonomous self-pollination in the event of pollinator failure. Another innovation is that flowers provide an arena for extended pollen-tube competition, which may promote offspring fitness by choice of superior sporophytic fathers (Mulcahy 1979; Skogsmyr and Lankinen 2002; Pannell and Labouche

2013). Pollen competition may also screen out genetically inferior male gametophytes, which potentially reduces inbreeding depression and might promote the persistence of mixed mating systems (Armbruster and Rogers 2004; Goodwillie *et al.*2005; Lankinen and Armbruster 2007; cf. Igic and Busch 2013). The structures and secretions of angiosperm flowers also promote using animals for pollen and seed dispersal; this may have allowed populations to be more dispersed and achieve targeted colonization of favourable microenvironments (Stebbins 1974; Regal 1977; Crepet and Niklas 2009). All of these features should lead to greater population persistence and hence lower extinction rates.

Another advantage of angiosperm flowers is that they are modular units comprising many parts of differing degrees of integration. As modules, they can maintain some degree of phenotypic independence from variation in the rest of the plant, which is adaptive in the face of pollinator-mediated stabilizing selection for flower size and shape (Berg 1960;Armbruster *et al.* 1999, 2004; Hansen *et al.* 2007; Pélabon *et al.* 2011;Murren 2012). Floral parts themselves range from nearly independent to highly integrated statistically (Conner and Via 1993; Conner and Sterling 1995, 1996; Armbruster *et al.* 1999, 2004; Herrera 2001; Herrera *et al.*2002; Anderson and Busch 2006; Hansen *et al.* 2007; Ordano *et al.* 2008;Pélabon *et al.* 2011; Alcantara *et al.* 2013) or structurally/developmentally (e.g. Armbruster *et al.* 1994, 2004). This variation in degree of integration allows tremendous adaptability in response to selection by different pollinators. For example, if sepals are under different selection than petals, their quasi-independence allows different evolutionary responses (e.g. Armbruster *et al.* 1999). If styles and stamens, for example, are selected to be of the same length (integration, in the face of variation), they can do so even though other floral traits experience different selective pressures (Conner and Via 1993; Conner and Sterling 1995,1996). Numerous semi-independent floral parts also allow the evolution of a diversity of complex interactions with pollinators, and, in fruit, seed dispersers (Stebbins 1974).

The above arguments imply reduced extinction rates for flowering plants compared with gymnosperms. The rapidity of floral development presumably also enabled them to occupy new environments, such as highly seasonal and successional habitats (including post-fire sites) requiring rapid growth and quick reproduction. Such habitats were thought to have become more abundant during the early and mid-

Cretaceous, when much angiosperm evolution was occurring (Axelrod 1970; Raven and Axelrod 1974; Stebbins 1974; Lamont and He 2012). Flower structure and complexity as described above suggest further that flowering plants had the potential to radiate by use of different pollinators and seed dispersers (see Evolution of floral disparity, below).

The above features lead to potentially rapid and labile evolutionary response of flowers to divergent selective pressures. This allows for increasing floral disparity between species in sympatry and species in allopatry but later in secondary sympatry. This capacity to diverge and specialize may thus also reduce extinction rates because it allows the compatible packing of more species (with narrower pollination niches) into communities (i.e. with minimal reproductive costs incurred by sharing pollinators).

Increased Speciation Rates

A classical explanation for high angiosperm diversity is that specialized flower–pollinator relationships increase speciation rates. Thus, in addition to reducing extinction rates and increasing the opportunity for the evolution of disparity (see below), flowers with the right combination of traits may have contributed to increased speciation rates in the angiosperms. Arguments and evidence for this relationship come from three sources: (i) classical floral-isolation models, (ii) phylogenetic-comparative analyses, and (iii) ecological genetic experiments.

Classical Floral Isolation Model

As noted above, Grant (1949, 1971, 1994*a*) recognized the potential importance of certain kinds of flowers in increasing speciation rates through establishment of pre-zygotic isolating mechanisms, specifically through differences in pollination ecology (floral isolation *ss.*). Complex flowers may attract only certain species of pollinators, which could lead to floral isolation from related species attracting other pollinator species. This form of ethological isolation is based on traditional ideas of specialization in flower–pollinator relationships, as captured in older syndrome literature (see Faegri and van der Pijl 1979). Some relationships between plants and their pollinators are sufficiently specialized that ethological isolation may influence speciation and reinforcement (e.g. in sexually deceptive orchids and plants pollinated by male-euglossine bees;Dressler 1968; Cozzolino and Widmer 2005; Pansarin and do Amaral 2009; but see Armbruster *et al.* 1992), but this is almost certainly the exception not the rule in angiosperms (Waser 1998, 2001; Armbruster and Muchhala 2009).

Some of the most intriguing data on ethological isolation in orchids come from comparisons of food-deceptive and sexually deceptive orchids (Scopece *et al.* 2007; Cozzolino and Scopece 2008). Generally, sexually deceptive orchids have strong pre-mating isolation and weak post-mating barriers, while food-deceptive orchids have strong post-mating isolation and weak pre-mating barriers. Sexual deception reflects unusual specialization based on chemical exploitation of one or a few pollinator species, whereas pollination in food-deceptive orchids is less specialized, attracting a range of pollinators much as do food-reward flowers. Interestingly, post-mating isolation has evolved in a clock-like manner, whereas pre-mating barriers have not, supporting the idea that, for angiosperms with food-deception and food-reward pollination systems, speciation commonly occurs through gradual divergence in allopatry (Cozzolino *et al.* 2005; Cozzolino and Scopece 2008).

Another form of ethological isolation operates through flower constancy, the tendency of individuals of some species of animals to be temporarily faithful to a single flower species or morph (Grant 1950; Waser 1986). There are several possible reasons that animals (mostly bees) may adopt constancy as a foraging strategy, but the consensus is that it is a way to reduce handling time and possibly search efficiency. It can be expected that floral features that increase a bee's learning time for handling will promote greater constancy, although supporting data are sparse and weak (Chittka *et al.* 1999; Armbruster *et al.* 2014*b*). If this is the case, however, it could result in higher speciation rates in complex flowers and hence explain greater clade diversity (e.g. Sargent 2004; Kay and Sargent 2009). However, constancy is unlikely to be a sole mechanism of reproductive isolation between incipient species in sympatry. This is because it is rare for any animal to be perfectly constant. For example, bumble bees, which are often highly constant, still make occasional (to numerous) interspecific-flower transitions (Heinrich 1976). Euglossine bees can be inconstant when flower handling is similar across plant species (e.g.Armbruster and Herzig 1984). Indeed Chittka *et al.* (1999) state: '...there is good evidence against the notion that pollinator constancy is involved in speciation or maintenance of plant species integrity'. Nevertheless, there are reasons to expect that plants will have evolved constancy-promoting floral features, because these increase reproductive fitness (dispersing more pollen to conspecific stigma and receiving more conspecific pollen), even though these rarely play a role in reproductive isolation. Instead, because floral features increasing constancy probably play a role in enhancing reproductive success, they may have increased diversification rates by reducing extinction rates (see above).

Certain floral traits may increase speciation rates through enhancement of mechanical isolation (Grant 1949, 1971, 1994*a*), where pollinators are restricted to a subset of visitors by the specialized fit of flowers to 'preferred' pollinators, or where flowers place pollen in a specific place on the pollinator (with stigma contact in the same location) not used by other species. Early studies of orchid speciation illustrate this concept nicely. As noted above, if two male-euglossine pollinated orchids produce chemically different fragrances, they are ethologically isolated (different bees attracted; Dressler 1968; Pansarin and do Amaral 2009). When visitors overlap, however, often related species diverge in the size of the flower and the size of the effective pollinator (mechanical isolation;Dodson 1962; Dressler 1968). In other cases, related orchids share visitors but partition where on the pollinator pollinaria are placed (mechanical isolation; Dressler 1968).

Phylogenetic Evidence

Under the assumption that greater clade species richness reflects higher speciation rates, several authors have used phylogenetic data to suggest that animal pollination increases speciation rates over abiotic pollination (Dodd *et al.* 1999) or that plants with more specialized pollination have higher speciation rates than those with less specialized pollination (Hodges and Arnold 1995; Hodges 1997; Sargent 2004; Kay and Sargent 2009; Schiestl and Schlueter 2009). Of course, most of these authors acknowledge that lower extinction rates could also have played a role in this association (Armbruster and Muchhala 2009; Kay and Sargent 2009).

Ecological Genetic Experiments

A series of elegant experiments with two *Mimulus* species (Phrymaceae),*M. cardinalis* and *M. lewisii*, has shown the genetic basis of floral signals determining flower choice by pollinators (Schemske and Bradshaw 1999;Bradshaw and Schemske 2003; Ramsey *et al.* 2003). This approach is very powerful, although it was not possible to show complete isolation being generated by the detected genetic differences (see Waser 2001). Similar studies identifying the genetic basis of differences in pollinator attraction between related plant species have now been conducted in several other lineages, e.g. *Ipomopsis* (Polemoniaceae; Nakazato *et al.* 2013), *Iris*(Iridaceae; Brothers *et al.* 2013), and *Petunia* (Solanaceae; Hermann *et al.*2013). However, it remains a significant challenge to distinguish between the role of floral signals in generating or maintaining reproductive isolation vs. simply improving reproductive fitness in sympatry. This challenge plagues all the approaches reviewed here.

Problems with the Enhanced Speciation Model

It seems clear that floral specialization is often associated with clade success (i.e. species richness; Hodges and Arnold 1995; Hodges 1997;Sargent 2004; Kay and Sargent 2009). However, it is less clear what causes this association. Armbruster and Muchhala (2009) laid out several possible causes of the association between floral specialization and clade species richness (see Fig. 2), and suggested several lines of evidence that could be used to distinguish between them. Data to date are sparse, although circumstantial cases have been made. Given that only a tiny amount of inter-morph pollination will swamp any divergence except that driven by very strong selection (Wright 1951; Roughgarden 1979; Waser 2001;Armbruster and Muchhala 2009), it is important to assess the degree of reproductive isolation that can result from different kinds of floral specialization. It is not clear that the fidelity of pollinator species and individuals visiting specialized flowers (ethological isolation; cf. Waser 1998, 2001; Chittka *et al.* 1999) is great enough to preclude the minimal pollen flow needed to swamp divergence. It seems even less likely that pollen placement and stigma contact locations usually differ between related species (mechanical isolation; Grant 1994*a*) sufficiently to preclude inter-morph pollinations swamping genetic divergence (Waser 2001;Armbruster *et al.* 2014*b*; Armbruster and Muchhala 2009).

Waser (1998, 2001) pointed out that most flowers (at least in the North Temperate Zone; cf. Johnson and Steiner 2000; Armbruster 2006) have fairly generalized pollination and are visited by several to many pollinator species. It would therefore be difficult for ethological isolation at the pollinator-species level to be complete enough to provide more than a limited degree of assortative mating. Thus, ethological isolation may often enhance reproductive fitness in sympatry, but it is unlikely to maintain species 'integrity'. Although Kay and Sargent (2009) and many others have suggested it acts multiplicatively along with other isolating mechanisms to generate complete reproductive isolation, the effect of differential attraction of pollinators alone seems too weak to be responsible for raising speciation rates in most cases (euglossine-pollinated and sexually deceptive orchids discussed above are possible exceptions).

EVOLUTION OF PHENOTYPIC DISPARITY

As noted above, phenotypic disparity refers to the phenotypic variation exhibited among related species (also called 'morphological' or 'phenotypic diversity'; Foote 1997). Speciation is usually associated with disparity, but clades vary dramatically in terms of how much disparity is generated with speciation. Pollinator shifts can be viewed as ecological disparity (a component of phenotypic disparity), which may or may not be associated with morphological disparity or speciation. The evolutionary timing of disparity is something that has received attention in the literature on animals and fossils but has not been addressed, as far as I know, in the phylogenetic-comparative literature on plant species. The several studies of animals suggest that the rate of increase in phenotypic disparity rises early in most adaptive radiations and then declines (e.g.Lumbsch *et al.* 2010; Hughes *et al.* 2013). It remains to be established whether this is a general trend or if it is true for flower traits.

Pollinator-Mediated Divergence and Pollinator Shifts

The diversity of extant pollinators and associated floral phenotypes, as reflected in pollination syndromes, attests to the importance of pollinator-mediated adaptive divergence of flowering plants (Fenster *et al.* 2004). However, the level and mechanisms by which this divergence occurs cannot be readily inferred from such broad flower–pollinator associations. Most insights into the possible processes of floral divergence come from studies of pollination ecotypes or of closely related species (e.g. congeners) in a phylogenetic context.

Wilson and Thomson (1996) recognized five processes potentially increasing floral disparity (divergence): (i) adaptation to distinct pollination niches, (ii) character displacement, (iii) adaptive wandering (where temporal variation in selective pressures can result in divergence without an overall difference in net selection), (iv) character correlations (where selection on one trait causes correlated response in another), and (v) genetic drift. These processes are detected primarily by comparing populations or species, which I review briefly below.

Armbruster and colleagues (Armbruster 1993; Armbruster and Muchhala 2009) have tried to classify and quantify the types of shifts in pollination systems. We recognized three types of shifts: (i) quantitative shifts, where transitions occur through small cumulative changes in quantitative traits in response to selection mediated by quantitative shifts in importance of different pollinators; (ii) qualitative shifts (e.g. colour, reward chemistry) with an intermediate phase when both old and new pollinators are present and effective; and (iii) qualitative shifts without an intermediate phase, where a qualitative change in floral features results immediately in a new pollinator. The first and second shifts are consistent with Stebbins' (1970,1974) 'gradualistic principle', whereas the third is consistent with instantaneous speciation (see Divergence of species). While some detailed genetic studies appear to support the instantaneous pollinator-shift model (Schemske and Bradshaw 1999; Bradshaw and Schemske 2003), more recent genetic studies suggest incremental multi-locus change, supporting the gradual-divergence model (Dell'Olivo and Kuhlemeier 2013; Nakazato *et al.* 2013). Below I review additional evidence for, and criticisms of, these models.

Divergence of Populations

Comparing the pollination biology of conspecific populations can reveal divergent selective pressures, the origins of pollination ecotypes and character displacement.

Divergent Selective Pressures

Galen and collaborators were among the first to show local variation in phenotypic selection apparently generating pollination ecotypes; divergent selection was associated with genetic and floral-morphological differentiation between *Polemonium* (Polemoniaceae) populations in the alpine and subalpine (Galen and Newport 1988; Galen 1989; Galen *et al.*1987, 1991). A similar, recent study also found differences in selection and floral morphology in alpine and subalpine *Trollius* (Ranunculaceae;Zhao and Huang 2013). Subsequent to the landmark study by Galen and her collaborators, a number of additional studies have documented spatial variation in phenotypic selection gradients; however, very few have documented differences in selection gradients consistent with patterns of phenotypic divergence (see the discussion in Maad *et al.* 2013). In the absence of this information, it is difficult to ascertain if spatial variation in selection gradients detected in any one study year actually represents long-term spatial differences in selection, given the extent of year-to-year variation in selection gradients (e.g. Schemske and Horvitz 1989; Parra-Tabla and Vargas 2004; Rees and Ellner 2009; Morales *et al.* 2010; seeSiepielski *et al.* 2009). One way to increase confidence in detecting evolutionarily meaningful divergent selection is to modify potential selection experimentally, as Sletvold *et al.* (2013) have done by manipulating both pollination and the height of the graminoid/herbaceous vegetation surrounding the orchid flowers under study; they showed that variation in the height of surrounding vegetation can generate significant variation in pollinator-mediated selection on floral traits. Ehrlen *et al.*(2002) and Sletvold *et al.* (2010) provide examples of another important experimental manipulation in floral phenotypic-selection studies: comparing plants subjected to both manual and natural-pollination treatments to identify phenotypic selection mediated by pollinators.

Pollination Ecotypes

Ecotypic divergence results from genetic response to divergent selection on conspecific populations that are geographically separated (usually reproductively isolated by distance). Pollination ecotypes are conspecific populations of plants that have diverged in pollination ecology, where that divergence has a genetic basis. (For examples of pollination ecotypes in addition to those discussed below, see the special issue of *Annals of Botany* 113(2) (2014).) Selective drivers of this divergence may be direct differences in abundance, reliability or behaviour of different pollinator species, or indirect differences in pollinator availability imposed by competing flower species. Genetic divergence across plant populations is reflected in differences in pollination ecology, including (i) divergence in pollinator species attracted as a result of different floral morphology and/or chemistry (rewards, advertisement colour or fragrance), (ii) divergence in which subset of floral visitors is used as pollinators, often as a result of differences in flower size, (iii) divergence in where anthers and stigmas contact the pollinator (without divergence in pollinator species), and/ or (iv) divergence in time of day or season that flowers are receptive for pollination.

Most studies of pollination ecotypes have considered divergence in the pollinator species attracted. For example, Breedlove (1969) showed that a species of *Fuchsia* (Onagraceae) in Chiapas, Mexico, attracted and was pollinated by both bees and hummingbirds in most sites, but where sympatric with another species of *Fuchsia*, the first species specialized on attracting hummingbirds. This is an example not only of ecotypic differentiation but also of character displacement. Whalen (1978) showed character displacement and ecotypic differentiation in flower size and pollinators and in populations of Mexican *Solanum* (Solanaceae). Miller (1981) documented among-population variation in flower colour and nectar-spur length in *Aquilegia caerulea*; the variation appeared to be associated with differences in pollinating hawk-moth faunas. Inoue (1983)described pollination ecotypes in *Platanthera* (Orchidaceae) in Japan.Armbruster (1985) found a similar pattern of differential attraction of pollinators of different sizes to populations of *Dalechampia scandens*producing different amounts of reward. Robertson and Wyatt (1990) found that populations of *Plantanthera ciliaris* (Orchidaceae) in the Appalachian Mountains and coastal plain of the southeastern USA were pollinated by different *Papilio* (swallowtail) species, and these had different proboscis lengths; the orchid populations (ecotypes) diverged in nectar-spur length in the direction that matched the difference in pollinator proboscis lengths. Galen *et al.* (1991) documented ecotypic divergence in floral morphology and volatile chemistry in *P. viscosum* in the Rocky Mountains, resulting in attraction of bumble-bee pollinators in alpine populations and fly pollinators in subalpine populations.

Since these early studies of pollination ecotypes attracting different pollinators, the field has grown markedly. Johnson, Anderson and collaborators have documented pollination ecotypes involving phenotypic divergence and attraction of different pollinators to conspecific populations of *Gladiolus* (Iridaceae) and several orchids in South Africa (e.g. Johnson 1997, 2010; Johnson and Steiner 1997; Anderson *et al.*2010; Peter and Johnson 2014). Valiente-Banuet *et al.* (2004) found that populations of *Pachycereus* cacti in tropical Mexico specialized on resident bat pollinators (flowers open and nectar secretion only at night), whereas more northerly subtropical populations had more generalized pollination involving diurnal insects, probably birds and migratory bats (flowers open and nectar secretion both during day and night). Arroyo, Pérez-Barrales and collaborators found that geographic variation in floral traits and phenotypic integration of *Narcissus papyraceus* flowers was associated with differences in pollinating faunas (Pérez-Barrales *et al.* 2007; Santos-Gally *et al.* 2013). In a classical reciprocal-transplants study of two*Platanthera* orchid ecotypes in Sweden, Boberg *et al.* (2014) showed that the long-spurred forest ecotype had higher reproductive success in the forest environment than did the meadow ecotype (although the reverse transplant experiment did not detect any difference), consistent with the differences in proboscis length of the main pollinators, and as expected from Darwin's hypothesis of spur-proboscis-length coevolution. Finally, clinal ecotypic variation in floral morphology and associated pollinator faunas have been reported for South American *Nicotiana* (Solanaceae) in Argentina (Nattero *et al.* 2011) and *Campanula* (Campanulaceae) in Norway (Maad *et al.* 2013).

Another axis of ecotypic divergence is using different subsets of visitors as pollinators. For example in *Dalechampia*, large resin-gland-to-stigma distances preclude pollination by small resin-collecting bees, even though they may visit and obtain rewards (Armbruster 1988).

Dalechampia scandens ecotypes vary in gland–stigma distances, so some ecotypes utilize small bees as pollinators and others utilize only large bees (Armbruster 1985). A similar example is that of euglossine-pollinated orchids employing only a subset of the bees attracted as pollinators (Dodson 1962; Dressler 1968), although the observed differences here occur at the level of closely related species. Ecotypic or subspecies variation in flower size in plants is common (e.g. Baldwin *et al.* 2011;Pettengill and Moeller 2012; Maad *et al.* 2013), and in many cases this probably reflects similar patterns of divergence in utilization of pollinators; however, the drivers of such differentiation remain to be described in this context, being more often attributed to mating-system differences (e.g. Armbruster *et al.* 2002; Elle and Carney 2003; Pettengill and Moeller 2012).

A third axis of ecotypic divergence is in the location of pollen placement on, and stigma contact with, pollinators. Grant (1994*a*, *b*) discussed this in the context of interspecific reproductive isolation, and it can be expected to be common at the level of ecotypes. However, there are only a few documented examples. One is reported from southwestern Australian*Stylidium* (Stylidiaceae), where conspecific populations have diverged from one another such that they overlap less with sympatric congeners in location of stigma and anther contact with pollinators than would be expected by chance (Armbruster *et al.* 1994). A similar situation has been documented in Andean *Burmeistera* (Campanulaceae) by Muchhala and Potts (2007).

The fourth axis of potential ecotypic divergence is in flowering time. There is some evidence that ecotypes can diverge in the time of day or season that flowers are receptive for pollination. For example, pollination ecotypes of *D. scandens* tend to differ from each other in the time of day that their blossoms open, and this again follows a character-displacement pattern from sympatric congeners (although not statistically significant;Armbruster and Herzig 1984; Armbruster 1985). Interspecific variation in time of anthesis and pollen release in both African and neotropical *Acacia*sl. (Stone *et al.* 1998; Raine *et al.* 2007) suggests another system to investigate for ecotypic divergence. However, the large constraints (phylogenetic signal) on seasonal variation in flowering time (Kochmer and Handel 1986; Davies *et al.* 2013) suggest that ecotypic divergence in flowering phenology may be less common, although there are some good examples (see Anderson *et al.* 2010; Peter and Johnson 2014). Note that seasonal flowering-time divergence of species has a long history of study in the context of community assembly and species coexistence (e.g. Stiles 1975; see review in Rathcke and Lacey 1985).

Divergence of Species

Most insights into major shifts in pollination ecology come from comparing related species. Modern studies have based such comparisons on phylogenetic information, now mostly molecular phylogenies. The main issue considered is how and why shifts in pollination systems occur. To address this question, it is useful to differentiate the kinds of shifts that can occur. Stebbins (1974) argued that all shifts between pollination systems occurred during an intermediate phase, during which both old and new pollinators were effective. Stebbins' argument here is that, for natural selection to be important in switches between pollinators, there must be a series of variable intermediate phenotypes for it to act upon. If the old pollinator is lost immediately upon the new one being gained (resulting in reproductive isolation), then only 'species selection' (differential extinction and/or speciation of lineages) 'chooses' between the two new lineages (Stanley 1975; Gould 2002; Rabosky and McCune 2010; see also Fernández-Mazuecos *et al.* 2013).

Stebbins' gradualistic view has been countered by work on orchids (Dodson 1962; Dressler 1968; Schiestl and Ayasse 2002; Ayasse *et al.*2011; Vereecken *et al.* 2011; but see Bradshaw *et al.* 2010), and more recently work on the molecular basis of pollinator discrimination (Schemske and Bradshaw 1999; Bradshaw and Schemske 2003; Yuan *et al.*2013). These studies indicate that small genetic changes might sometimes lead to 'instantaneous speciation'.

Studies of Congeneric Divergence

Most evidence for how shifts in pollination systems occur comes from study congeneric species. The number of such studies has grown tremendously in recent years and include Armbruster (1993; *Dalechampia*, Euphorbiaceae), Hapeman and Inoue (1997; *Platanthera*, Orchidaceae),Baum *et al.* (1998; *Adansonia*, Malvacae/Bombacaceae), Johnson *et al.*(1998; *Disa*, Orchidaceae), Kay *et al.* (2005; *Costus*, Costaceae), Perez *et al.* (2006; *Schizanthus*, Solanaceae) and Wilson *et al.* (2007; *Penstemon*, Plantaginaceae). Sometimes additional questions are addressed using data on the evolution of pollination systems. For example, Armbruster and Baldwin (1998) used *Dalechampia*, and Tripp and Manos (2008) used*Ruellia* (Acanthaceae) as comparative study systems for assessing whether or not specialization pollination can transition to generalized pollination (see also Martens-Rodriguez *et al.* 2010). Ley and Classen-Bockhoff (2009) commented on the role of intermediate forms in pollinator shifts in African Marantaceae. Smith *et al.* (2008), Martens-Rodriguez *et al.* (2010)and Sakai *et al.* (2013) examined the association between transitions in floral traits and functional groups of pollinators of flowers of *Iochroma*(Solanaceeae), Gesnerieae (Gesneriaceae) and Bornean gingers (Zingiberaceae), respectively, finding that some evolving floral traits were tightly associated with transitions between pollinators while others were not (see also Thomson and Wilson 2008). Friedman and Barrett (2008) conducted similar analyses of both order and direction of change of traits associated with shifts from animal to wind pollination across the angiosperms.

Very few studies have attempted to classify or tally the kinds of shifts (as described above) between pollination systems. Only two studies of which I am aware have tallied the frequencies of different types of pollinator shifts. The first is of *Dalechampia* (Armbruster 1993). Here 14 reconstructed shifts between pollinator systems were identified as quantitative, 3–6 as qualitative with the intermediate phase and 1–3 were identified as qualitative without the intermediate phase. The second study is of the Bignonieae (Bignoniaceae; Alcantara and Lohmann 2010), wherein the authors described 10 shifts with an intermediate phase and 19 without the intermediate phase (although this is based on floral morphology not pollinator observations). The first study largely supported Stebbins' gradualistic principle, but as a trend not a rule; i.e. most shifts had an intermediate stage during which natural selection could act to push or reverse the shift. Importantly, however, at least one shift appeared to lack any intermediate phase, and this could lead to instantaneous, non-adaptive speciation (by a single mutation affecting fragrance chemistry, leading to attraction of male euglossine bees instead of females). The second study suggested that shifts not involving the intermediate phase are much more common than Stebbins thought, although there is some uncertainty in this conclusion due to the lack of direct pollination data. The second study suggests an important role for non-adaptive speciation, at least in the Bignonieae. However, because all estimated phylogenies are incomplete representations of the true evolutionary history, these and other trends noted in this paper need to be assessed in many additional studies before placing too much confidence in any broad conclusions.

Increasing Disparity without Pollinator Shifts

Not all increases in phenotypic disparity are associated with shifts in pollinators. For example, in *Dalechampia* over half (15) of 29 reconstructed speciation events associated with increased quantitative phenotypic disparity (morphological divergence) involved no change in pollinators (reanalysed from data in Armbruster 1993). Ellis and Anderson (2012) reviewed the topic recently and concluded that phenotypic divergence in the absence of pollinator change can come about in several ways. Behavioural variation in a single pollinator species can impose divergent selective pressures on plant species and drive adaptive divergence. Similarly, regional variation in the models that non-rewarding flowers mimic can generate disparity even though the same pollinator species is employed (Ellis and Anderson 2012). Variation in non-pollinator selective agents (e.g. florivores, seed predators) can also drive the evolution of disparity even when pollinators do not differ (e.g. Pérez-Barrales *et al.* 2013; see the section above on conflicting selection).

When related species share pollinators in sympatry, they may often diverge phenotypically, as, for example, the morphological diversity seen in*Pedicularis*, even though nearly all are pollinated by similar bumble bees (Macior 1983; Grant 1994*b*; Eaton *et al.* 2012; Huang and Shi 2013;Armbruster *et al.* 2014*b*). Similarly, floral disparity in *Stylidium*(Stylidiaceae) and *Burmeistera* (Campanulaceae) seems to be associated with divergence in sites of pollen placement on the same pollinator taxa instead of using different pollinator species (Armbruster *et al.* 1994, 2004,2009*a*; Muchhala and Potts 2007).

SPECIATION AND MACROEVOLUTION

One of the major evolutionary discussions in the late 20th century was whether macroevolution was merely an extension of microevolution over longer time periods, or whether macroevolution was, to some extent, decoupled from microevolution (e.g. Gould 1980). Although the debate has died down, there has not really been any resolution, with paleobiologists still tending to favour decoupling and neontologists favouring lack of decoupling. If speciation is largely adaptive, then it forms a link between micro- and macroevolution and the two are coupled to some considerable extent. However, if speciation is largely non-adaptive, then micro- and macroevolution are largely decoupled (Stanley 1975;Gould 2002).

It is thus useful to understand more about how speciation occurs in flowering plants. Whenever flowering-plant speciation is accompanied by genetically based quantitative shifts in pollinators or by qualitative shifts in pollinators with an intermediate phase, then the speciation even is potentially adaptive, as in ecological speciation (Nosil 2012). Thus the relative frequency of instantaneous (non-adaptive) pollinator shifts, such as between sexually deceptive systems (e.g. floral volatiles mimicking insect sex pheromones) or between euglossine pollinators, relative to the other two forms gives us insights into the degree to which macroevolution might be decoupled from microevolution. The limited surveys to date suggest that both potentially adaptive (quantitative and qualitative with an intermediate phase) and potentially non-adaptive (without an intermediate phase) pollinator shifts occur, and the latter could lead to instantaneous speciation in some cases. However, overall, it appears that adaptive shifts are probably more common, suggesting potentially strong links between micro- and macroevolutionary patterns and processes. Note, however, that the potential role of natural selection in gradual shifts between pollinators does not, by itself, demonstrate its importance.

DISCUSSION

The rapid increase of species diversity (i.e. the increase of clade species richness through speciation) with increasing phenotypic disparity (phenotypic diversity) is usually termed an 'adaptive radiation' (Schluter 2000). The accumulation of disparity with species diversity may drive further adaptive speciation as a positive feedback. Any feature that increases the likelihood of reproductive isolation may thus have major effects on both diversification and phenotypic disparity. Features that can respond to selection for reinforcement (increased reproductive isolation as a result of selection on partially inter-fertile species in sympatry; but seeMoyle *et al.* 2004) or selection for reproductive character displacement (selection against inter-mating in the absence of inter-fertility) thus may play a disproportionate role in ramping up clade species richness and phenotypic disparity (but see Rabosky and Matute 2013 for two animal counter-examples).

It is important to remember that there are strong fitness costs to interspecific pollination, even if species are completely inter-sterile and fully reproductively isolated (e.g. Muchhala *et al.* 2010). Further, unlike in reproductive isolation, only a small percentage improvement in getting pollen to, and from, the right species can be biologically significant and selected for. In contrast, for initial reproductive isolation or reinforcement, such small improvements would usually be genetically inconsequential.

It is also important to remember that most mechanical 'isolation' in plants is probably too incomplete to be of any biological significance in reproductive isolation (Armbruster and Muchhala 2009; Armbruster *et al.*2014*b*), except in plants with pollinia, like orchids, or in combination with other isolating factors (Kay and Sargent 2009). However, placing pollen in different average locations on shared pollinators, even with overlap, is biologically significant because it enhances the reproductive fitness of individuals.

Adaptive inaccuracy and its components, maladaptive bias and adaptive imprecision, provide useful insights into the fitness costs and benefits of floral features, such as herkogamy (Armbruster *et al.* 2009*a*, *b*) and differences between sympatric species sharing pollinators (e.g. Armbruster*et al.* 2014*b*). In the few species that have been studied, low adaptive accuracies are generated primarily by low precision, especially in pollen placement (Armbruster and Muchhala 2009; Armbruster *et al.* 2009*a*,2014*a*). Imprecision increases dramatically when accounting for realized inaccuracy, even in species with highly specialized flowers (Armbruster *et al.* 2014*b*). For example, large increases in imprecision (from fundamental to realized) have been observed in *Stylidium* (Stylidiaceae), where the increase is caused primarily by variation in pollinator orientation on landing on the flowers and/or pollen redistribution after the visit (W. S. Armbruster and J. A.Wege, unpubl. res.). These observations underscore the difficulty flowers with granular pollen have in achieving significant mechanical isolation by placing pollen in different locations on the same pollinators.

Prospects for Further Research

Despite a long history of research, the evolution of floral specialization, plant–pollinator interactions and angiosperm diversity remain fruitful areas of investigation, indeed ones that have grown dramatically in recent years (Fig. 6). One reason that this area of research has experienced an explosion in activity in the past few decades, and a reason for optimism about the future vibrancy of the field, is that the interaction between plants and pollinators is a measurable phenomenon that is tightly linked to reproductive fitness. Thus both ecological and comparative studies have the capacity to address challenging, fundamental evolutionary questions, as well as questions that have practical applications in the conservation of biological diversity and the assessment of the impacts of global change on ecosystem services (e.g. pollination). Nevertheless, this review has probably identified more uncertainties than well-established knowledge.

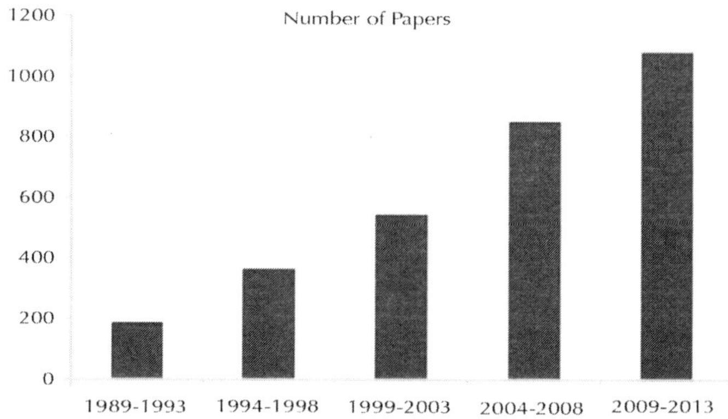

Figure 6: Number of papers on evolutionary aspects of pollination published in 5-year periods from 1989 to 2013. Data from Thomson ISI Web of Science, searching on "evolution AND pollination" on 30 Oct. 2013. Note that papers published prior to 1988 were excluded from the analysis because they did not include searchable abstracts.

A good understanding of fitness trade-offs in the interaction between floral features and pollinator effectiveness is critical to understanding the evolution of floral specialization. However, fitness trade-offs have been so rarely studied that we still do not know if they are common or uncommon. Especially rare are controlled experimental studies, which are the most unambiguous way to detect trade-offs. More studies using factorial experiments with different species of pollinators and different floral phenotypes would be valuable (see Wilson and Thomson 1996; Aigner 2001, 2006; Muchhala 2007). Another line of investigation not yet pursued is whether fitness trade-offs in pollination have influenced long-term clade success. Do lineages without fitness trade-offs have different diversification rates than those with trade-offs, e.g. because those without trade-offs have reduced extinction, or because those with trade-offs have higher speciation rates? It should therefore be a fruitful line of investigation to assess in well-studied groups the diversification consequences of the presence vs. absence of fitness trade-offs across lineages.

Much discussion of the role of pollinator constancy in the reproductive isolation of sympatric plant species has accumulated in the literature. Yet outside bumble and honey bees, we know very little about constancy in most pollinators. There are good reasons to think it may be facultative at most, and in many situations and organisms, including some bees, it may be low to non-existent (e.g. Armbruster and Herzig 1984; Raine and Chittka 2005; Pohl *et al.* 2011; Ellis and Johnson 2012; but see Goulson and Wright 1998). There is some evidence that floral complexity that increases the bee's learning time in achieving efficient handling promotes greater constancy (Corbet *et al.* unpublished; but see Wilson and Stine 1996). If so, this could contribute to higher speciation rates in complex flowers and hence explain greater clade diversity for lineages with complex or difficult-to-handle flowers (e.g. Sargent 2004; Kay and Sargent 2009).

To date there are very few phenotypic selection studies of ecotypes showing that contemporary selection maintains phenotypic differences between populations. This area deserves much more work. This is particularly challenging because year-to-year variation means that evolutionarily significant differences in selection may not be detected in any single year, even when they have been important in the phenotypic and genetic divergence of populations.

It remains a significant challenge to distinguish between the role of floral signals and morphology in generating or maintaining reproductive isolation vs. simply improving reproductive fitness in sympatry. The situation is not helped by the fact that 'floral isolation' is commonly used to refer to any degree of segregated pollen flow caused by flower or pollinator differences, even when it results in only a small percentage improvement. I fear that this has led some to think that reproductive isolation can commonly be generated by floral differences alone or in combination with other weak factors ('components of reproductive isolation'). Given that only a tiny amount of inter-morph pollination will swamp any divergence except that driven by strong selection (Wright 1951; Roughgarden 1979; Waser 2001; Armbruster and Muchhala 2009), it is important to assess the degree of reproductive isolation that can result from different kinds of floral specialization. One critical feature is the fidelity of pollinator species and individuals visiting specialized flowers (ethological isolation; cf. Waser 1998, 2001; Chittka *et al.* 1999) and the degree to which pollen placement and stigma contact in different locations precludes inter-morph pollination (mechanical isolation; Grant 1994*a*). However, when pre-pollination isolation is incomplete, but no hybrids are observed in the field, as may often be the case (Cozzolino *et al.* 2005;Cozzolino and Scopece

2008; Armbruster and Muchhala 2009), one has to suspect that post-pollination isolation is what really matters in the origins of species. Pre-pollination differences would then more likely reflect adaptation to sympatry (reinforcement or character displacement). However, all this being said, we still cannot make generalizations because we lack broad enough surveys of the association between hybridization and pre- vs. post-pollination isolation (Armbruster and Muchhala 2009; cf. Rabosky and Matute 2013).

The relationship between pollinator shifts and speciation remains an exciting line of investigation (see van der Niet and Johnson 2012;Dell'Olivo and Kuhlemeier 2013; Nakazato *et al.* 2013). New methods of estimating changes in diversification and speciation rates in lineages hold much promise for assessing the influence of pollinator shifts on rates of diversification (Armbruster *et al.* 2009*b*; Smith 2010). It might also be interesting to assess the association between different kinds of shifts and diversification. One early classification described, as discussed above, three kinds of pollinator shifts (quantitative, qualitative with an intermediate phase and qualitative shifts without an intermediate phase). This classification may need refining and certainly needs much more empirical testing. At the moment, however, it still seems a reasonable way to approach the process because of its direct connection to the causes and evolutionary consequences of pollinator shifts, potentially yielding insights into whether pollinator shifts (and speciation) are largely adaptive vs. non-adaptive and informing the micro–macroevolution discussion.

Also at the macroevolutionary scale, it will be interesting to explore the role of key innovations in the evolution of phenotypic disparity. At present, key innovations are investigated by looking at their putative effects on diversification rates. But of equal interest is the effect of a trait on both total disparity and lineage-corrected disparity (i.e. the morphological or ecological 'diversity' corrected for the number of lineages or species in the clade). Another line of investigation that may prove promising is to examine the evolutionary lability vs. conservatism of pollinator associations in relation to attraction traits (rewards, advertisements) and traits affecting floral–pollinator fit and pollen/stigma contact points. Is there a stronger phylogenetic signal in some features than others? Are sister species more likely to differ in one of these and be the same in the others? Is the evolution of certain features (e.g. reward chemistry) more constrained than others, and if so why? Does the evolutionary change in reward chemistry associated with pollinator shifts generally precede or lag behind evolutionary change in other traits?

Are diversity and floral disparity correlated in a non-trivial manner (beyond the sampling effect), and if so, how (see Rabosky *et al.* 2013)? Does speciation promote phenotypic evolution (Eldredge and Gould 1972;Gould 2002), or does floral evolvability promote speciation (Bolstad *et al.*2014; see also Rabosky *et al.* 2013)? Alternatively, might the evolvability of other aspects of phenotype (e.g. vegetative morphology, ion-uptake physiology) promote speciation, and increasing floral disparity reflect adaptive response to packing numerous sympatric species into the same 'pollination space' (see Armbruster *et al.* 1994; Armbruster and Muchhala 2009)?

SOURCES OF FUNDING

The research was supported by grants from the US National Science Foundation, the Norwegian Research Council, The British Council and the UK Royal Society.

ACKNOWLEDGEMENTS

I thank numerous students, postdocs and collaborators for help in the field and Marco Pautasso and two anonymous reviewers for helpful comments on this manuscript.

REFERENCES

1. Aigner PA. 2001. Optimality modeling and fitness trade-offs: when should plants become pollinator specialists? Oikos 95:177–184.
2. Aigner PA. 2004. Floral specialization without trade-offs: optimal corolla flare in contrasting pollination environments. Ecology 85: 2560–2569.
3. Aigner PA. 2006. The evolution of specialized floral phenotypes in a fine-grained pollination environment. In: Waser N, Ollerton

J, eds. Plant–pollinator interactions: from specialization to generalization, Chicago: University of Chicago Press, 23–46.

4. Alcantara S, Lohmann LG. 2010. Evolution of floral morphology and pollination system in Bignonieae (Bignoniaceae). American Journal of Botany 97:782–796.
5. Alcantara S, de Oliveira FB, Lohmann LG. 2013. Phenotypic integration in flowers of neotropical lianas: diversification of form with stasis of underlying patterns. Journal of Evolutionary Biology 26:2283–2296.
6. Anderson B, Alexandersson R, Johnson SD. 2010. Evolution and coexistence of pollination ecotypes in an African Gladiolus (Iridaceae). Evolution 64:960–972.
7. Anderson IA, Busch JW. 2006. Relaxed pollinator-mediated selection weakens floral integration in self-compatible taxa of Leavenworthia (Brassicaceae). American Journal of Botany 93:860–867.
8. Armbruster WS. 1985. Patterns of character divergence and the evolution of reproductive ecotypes of Dalechampia scandens (Euphorbiaceae). Evolution 39:733–752.
9. Armbruster WS. 1988. Multilevel comparative analysis of morphology, function, and evolution of Dalechampia blossoms. Ecology 69:1746–1761.
10. Armbruster WS. 1990. Estimating and testing the shapes of adaptive surfaces: the morphology and pollination of Dalechampia blossoms. American Naturalist 135:14–31.
11. Armbruster WS. 1991. Multilevel analyses of morphometric data from natural plant populations: insights into ontogenetic, genetic, and selective correlations in Dalechampia scandens. Evolution 45:1229–1244.
12. Armbruster WS. 1993. Evolution of plant pollination systems: hypotheses and tests with the neotropical vine Dalechampia. Evolution 47:1480–1505.
13. Armbruster WS. 2006. Evolutionary and ecological perspectives on specialization: from the arctic to the tropics. In: Waser N, Ollerton J, eds. Plant-pollinator interactions: from specialization to generalization. Chicago: University of Chicago Press, 260–282.

14. Armbruster WS, Baldwin BG. 1998. Switch from specialized to generalized pollination. Nature 394:632.
15. Armbruster WS, Herzig AL. 1984. Partitioning and sharing of pollinators by four sympatric species of Dalechampia (Euphorbiaceae) in Panama. Annals of the Missouri Botanical Garden 71:1–16.
16. Armbruster WS, Muchhala N. 2009. Associations between floral specialization and species diversity: cause, effect, or correlation? Evolutionary Ecology 23:159–179.
17. Armbruster WS, Rogers DG. 2004. Does pollen competition reduce the cost of inbreeding? American Journal of Botany 91: 1939–1943.
18. Armbruster WS, Herzig AL, Clausen TP. 1992. Pollination of two sympatric species of Dalechampia (Euphorbiaceae) in Suriname by male euglossine bees. American Journal of Botany 79: 1374–1381.
19. Armbruster WS, Edwards ME, Debevec EM. 1994. Character displacement generates assemblage structure of Western Australian triggerplants (Stylidium). Ecology 75:315–329.
20. Armbruster WS, Di Stilio VS, Tuxill JD, Flores TC, Velasquez Runk JL. 1999. Covariance and decoupling of floral and vegetative traits in nine neotropical plants: a reevaluation of Berg's correlationpleiades concept. American Journal of Botany 86:39–55.
21. Armbruster WS, Fenster CB, Dudash MR. 2000. Pollination 'principles' revisited: specialization, pollination syndromes, and the evolution of flowers. Det Norske Videnskapsacademi I. Matematisk Naturvidenskapelige Klasse Skrifter, Ny Serie 39:139–148.
22. Armbruster WS, Mulder CPH, Baldwin BG, Kalisz S, Wessa B, Nute H. 2002. Comparative analysis of late floral development and mating-system evolution in tribe Collinsieae (Scrophulariaceae, s.l.). American Journal of Botany 89:37–49.
23. Armbruster WS, Pe´labon C, Hansen TF, Mulder CPH. 2004. Floral integration and modularity: distinguishing complex adaptations from genetic constraints. In: Pigliucci M, Preston KA, eds. The evolutionary biology of complex phenotypes. Oxford, UK: Oxford University Press, 23–49.

24. Armbruster WS, Pe´rez-Barrales R, Arroyo J, Edwards ME, Vargas P. 2006. Three-dimensional reciprocity of floral morphs in wild flax (Linum suffruticosum): a new twist on heterostyly. New Phytologist 171:581–590.
25. Armbruster WS, Hansen TF, Pe´ labon C, Pe´rez-Barrales R, Maad J. 2009a. The adaptive accuracy of flowers: measurement and microevolutionary patterns. Annals of Botany 103:1529–1545.
26. Armbruster WS, Lee J, Baldwin BG. 2009b. Macroevolutionary patterns of defense and pollination in Dalechampia vines: adaptation, exaptation, and evolutionary novelty. Proceedings of the National Academy of Sciences of the USA 106:18085–18090.
27. Armbruster WS, Corbet SA, Vey AJM, Liu SJ, Huang SQ. 2014a. In the right place at the right time: Parnassia resolves the herkogamy dilemma by accurate repositioning of stamens and stigmas. Annals of Botany 113:97–103.
28. Armbruster WS, Shi X-Q, Huang S-Q. 2014b. Do specialized flowers promote reproductive isolation? Realized pollination accuracy of three sympatric Pedicularis species. Annals of Botany 113: 331–340.
29. Axelrod DI. 1970. Mesozoic paleogeography and early angiosperm history. Botanical Review 36:277–319.
30. Ayasse M, Stokl J, FranckeW. 2011. Chemical ecology and pollinatordriven speciation in sexually deceptive orchids. Phytochemistry 72:1667–1677.
31. Bakker FT, Culham A, Marais EM, Gibby M. 2005. Nested radiation in Cape Pelargonium. In: Bakker FT, Chatrou LW, Gravendeel B, Pelser PB, eds. Plant species-level systematics: new perspectives on pattern and process. Liechtenstein, Germany: ARG Gantner/ Lubrecht and Cramer, 75–100.
32. Baldwin BG, Kalisz S, Armbruster WS. 2011. Phylogenetic perspectives on diversification, biogeography, and floral evolution of Collinsia and Tonella (Plantaginaceae). American Journal of Botany 98:731–753.
33. Baum DA, Small RL, Wendel JF. 1998. Biogeography and floral evolution of baobabs (Adansonia, Bombacaceae) as inferred from multiple data sets. Systematic Biology 47:181–207.
34. Benevides CR, Evans DM, Gaglianone MC. 2013. Comparing the

structure and robustness of Passifloraceae - Floral visitor and true pollinator networks in a lowland Atlantic forest. Sociobiology 60: 295–305.

35. Berg RL. 1960. The ecological significance of correlation pleiades. Evolution 14:171–180.
36. Boberg E, Alexandersson R, Jonsson M, Maad J,A° gren J, Nilsson LA. 2014. Pollinator shifts and the evolution of spur length in the moth-pollinated orchid Platanthera bifolia. Annals of Botany 113:267–275.
37. Bolstad GH, Hansen TF, Pe´ labon C, Falahati-Anbaran M, Pe´rez- Barrales R, Armbruster WS. 2014. Genetic constraints and evolutionary divergence in Dalechampia blossoms. Philosophical Transactions of the Royal Society B (in press).
38. Bradshaw E, Rudall PJ, Devey DS, Thomas MM, Glover BJ, Bateman RM. 2010. Comparative labellum micromorphology of the sexually deceptive temperate orchid genus Ophrys: diverse epidermal cell types and multiple origins of structural colour. Botanical Journal of the Linnean Society 162:504–540.
39. Bradshaw HD, Schemske DW. 2003. Allele substitution at a flower colour locus produces a pollinator shift in monkeyflowers. Nature 426:176–178.
40. Breedlove DE. 1969. The systematics of Fuchsia section Encliandra (Onagraceae). University of California Publications in Botany 53: 1–69.
41. Brody AK. 1997. Effects of pollinators, herbivores, and seed predators on flowering phenology. Ecology 78:1624–1631.
42. Brothers AN, Barb JG, Ballerini ES, Drury DW, Knapp SJ, Arnold ML. 2013. Genetic architecture of floral traits in Iris hexagona and Iris fulva. Journal of Heredity 104:853–861.
43. Cariveau D, Irwin RE, Brody AK, Garcia-Mayeya LS, von der Ohe A. 2004. Direct and indirect effects of pollinators and seed predators to selection on plant and floral traits. Oikos 104:15–26.
44. Castellanos MC, Wilson P, Thomson JD. 2004. 'Anti-bee' and 'pro-bird' changes during the evolution of hummingbird pollination in Penstemon flowers. Journal of Evolutionary Biology 17:876–885.
45. Chittka L, Thomson JD, Waser NM. 1999. Flower constancy,

insect psychology, and plant evolution. Naturwissenschaften 86:361–377.

46. Conner J, Via S. 1993. Patterns of phenotypic and genetic correlations among morphological and life-history traits in wild radish, Raphanus raphanistrum. Evolution 47:704–711.
47. Conner JK, Sterling A. 1995. Testing hypotheses of functionalrelationships— a comparative survey of correlation patterns among floral traits in 5 insect-pollinated plants. American Journal of Botany 82:1399–1406.
48. Conner JK, Sterling A. 1996. Selection for independence of floral and vegetative traits: evidence from correlation patterns in five species. Canadian Journal of Botany 74:642–644.
49. Corbet SA. 1997. Role of pollinators in species preservation, conservation, ecosystem stability and genetic diversity. Acta Horticulturae 437:219–229.
50. Cozzolino S, Scopece G. 2008. Deceptive orchids: the promise of sex or food and its consequences for reproductive isolation. Philosophical Transaction of the Royal Society B 363:3037–3046.
51. Cozzolino S,Widmer A. 2005. Orchid diversity: an evolutionary consequence of deception? Trends in Ecology and Evolution 20:487–494.
52. Cozzolino S, Schiestl FP, Muller A, De Castro O, Nardella AM,Widmer A. 2005. Evidence for pollinator sharing in Mediterranean nectarmimic orchids: absence of premating barriers? Proceedings of the Royal Society B Biological Sciences 272:1271–1278.
53. Crepet WL, Niklas KJ. 2009. Darwin's second 'abominable mystery': why are there so many angiosperm species? American Journal of Botany 96:366–381.
54. Davies TJ, Wolkovich EM, Kraft NJB, Salamin N, Allen JM, Ault TR, Betancourt JL, Bolmgren K, Cleland EE, Cook BI, Crimmins TM, Mazer SJ, McCabe GJ, Pau S, Regetz J, Schwartz MD, Travers SE. 2013. Phylogenetic conservatism in plant phenology. Journal of Ecology 101:1520–1530.
55. Dell'Olivo A, Kuhlemeier C. 2013. Asymmetric effects of loss and gain of a floral trait on pollinator preference. Evolution 67:3023–3031.

56. Dieckmann U, Doebeli M, Metz JAJ, Tautz D (eds). 2004. Adaptive speciation. Cambridge, UK: Cambridge University Press.

57. Dodd ME, Silvertown J, Chase MW. 1999. Phylogenetic analysis of trait evolution and species diversity variation among angiosperm families. Evolution 53:732–744.

58. Dodson CH. 1962. The importance of pollination in the evolution of the orchids of Tropical America. American Orchid Society Bulletin 31:525–534, 641–649, 731–735.

59. Donoghue MJ. 2005. Key innovations, convergence, and success: macroevolutionary lessons from plant phylogeny. Paleobiology 31:77–93.

60. Dressler RL. 1968. Pollination by euglossine bees. Evolution 22: 202–210.

61. Eastman JM, Wegmann D, Leuenberger C, Harmon LJ. 2013. Simpsonian 'evolution by jumps' in an adaptive radiation of Anolis lizards. Cornell Univ. ArXiv Preprint, ArXiv 1305.4216 (http:// arxiv.org/abs/1305.4216).

62. Eaton DAR, Fenster CB, Hereford J, Huang S-Q, Ree RH. 2012. Floral diversity and community structure in Pedicularis (Orobanchaceae). Ecology 93:S182–S194.

63. Ehrlen J, Kack S, Agren J. 2002. Pollen limitation, seed predation and scape length in Primula farinosa. Oikos 97:45–51.

64. Eldredge N, Gould SJ. 1972. Punctuated equilibria: an alternative to phyletic gradualism. In: SchopfTJM, ed. Models in paleobiology. San Francisco: Freeman, Cooper & Co., 82–115.

65. Elle E, Carney R. 2003. Reproductive assurance varies with flower size in Collinsia parviflora (Scrophulariaceae). American Journal of Botany 90:888–896.

66. Ellis AG, Anderson B. 2012. Pollinator mediated floral divergence in the absence of pollinator shifts. In: Patiny S, ed. Evolution of plant-pollinator relationships. Cambridge, UK: Cambridge University Press, 237–262.

67. Ellis AG, Johnson SD. 2012. Lack of floral constancy by bee fly pollinators: implications for ethological isolation in an African daisy. Behavioral Ecology 23:729–734.

68. Faegri K, van der Pijl L. 1979. The principles of pollination ecology, 3rd ed. London: Pergamon.

69. Fenster CB, Armbruster WS, Wilson P, Dudash MR, Thomson JT. 2004. Pollination syndromes and floral specialization. Annual Review of Ecology, Evolution and Systematics 35:375–403.

70. Ferna´ndez-Marı´n H, Zimmerman JK, Wcislo WT. 2004. Ecological traits and evolutionary sequence of nest establishment in fungus-growing ants (Hymenoptera, Formicidae, Attini). Biological Journal of the Linnean Society 81:39–48.

71. Ferna´ndez-Mazuecos M, Blanco-Pastor JL, Gomez JM, Vargas P. 2013. Corolla morphology influences diversification rates in bifid toadflaxes (Linaria sect. Versicolores). Annals of Botany 112:1705–1722.

72. Findlay GP. 1978. Movement of column of Stylidium crassifolium as a function of temperature. Australian Journal of Plant Physiology 5: 477–484.

73. Foote M. 1997. The evolution of morphological diversity. Annual Review of Ecology and Systematics 28:129–152.

74. Freitas L. 2013. Concepts of pollinator performance: is a simple approach necessary to achieve a standardized terminology? Brazilian Journal of Botany 36:3–8.

75. Friedman J, Barrett SCH. 2008. A phylogenetic analysis of the evolution of wind pollination in the angiosperms. International Journal of Plant Sciences 169:49–58.

76. Galen C. 1989. Measuring pollinator-mediated selection on morphometric floral traits—bumblebees and the alpine sky pilot, Polemonium viscosum. Evolution 43:882–890.

77. Galen C. 1999. Flowers and enemies: predation by nectar-thieving ants in relation to variation in floral form of an alpine wildflower, Polemonium viscosum. Oikos 85:426–434.

78. Galen C. 2000. High and dry: Drought stress, sex-allocation trade-offs, and selection on flower size in the alpine wildflower Polemonium viscosum (Polemoniaceae). American Naturalist 156:72–83.

79. Galen C, Newport MEA. 1988. Pollination quality, seed set, and flower traits in Polemonium viscosum —complementary effects of variation in flower scent and size. American Journal of Botany 75: 900–905.

80. Galen C, Zimmer KA, Newport ME. 1987. Pollination in floral

scent morphs of Polemonium viscosum—a mechanism for disruptive selection on flower size. Evolution 41:599–606.

81. Galen C, Shore JS, Deyoe H. 1991. Ecotypic divergence in alpine Polemonium viscosum—genetic structure, quantitative variation, and local adaptation. Evolution 45:1218–1228.
82. Galen C, Kaczorowski R, Todd SL, Geib J, Raguso RA. 2011. Dosagedependent impacts of a floral volatile compound on pollinators, larcenists, and the potential for floral evolution in the alpine skypilot Polemonium viscosum. American Natutralist 177:258–272.
83. Gianoli E. 2004. Evolution of a climbing habit promotes diversification in flowering plants. Proceedings of the Royal Society B Biological Sciences 271:2011–2015.
84. Givnish TJ. 2010. Ecology of plant speciation. Taxon 59:1326–1366.
85. Goodwillie C, Kalisz S, Eckert CG. 2005. The evolutionary enigma of mixed mating systems in plants: occurrence, theoretical explanations, and empirical evidence. Annual Review of Ecology, Evolution and Systematics 11:15–39.
86. Gould SJ. 1980. Is a new and general theory of evolution emerging? Paleobiology 6:119–130.
87. Gould SJ. 2002. The structure of evolutionary theory. Cambridge, MA: Belknap Press.
88. Goulson D, Wright NP. 1998. Flower constancy in the hoverflies Episyrphus balteatus (Degeer) and Syrphus ribesii (L.) (Syrphidae). Behavioral Ecology 9:213–219.
89. Grant PR. 1972. Convergent and divergent character displacement. Biological Journal of the Linnean Society 4:39–68. Grant V. 1949. Pollination systems as isolating mechanisms in flowering plants. Evolution 3:82–97.
90. Grant V. 1950. The flower constancy of bees. Botanical Review 16: 379–398.
91. Grant V. 1966. The selective origin of incompatibility barriers in the plant genus Gilia. American Naturalist 100:99–118.
92. Grant V. 1971. Plant speciation. New York: Columbia University Press.
93. Grant V. 1994a. Modes and origins of mechanical and ethological

isolation in angiosperms. Proceedings of the National Academy of Sciences of the USA 91:3–10.

94. Grant V. 1994b. Mechanical and ethological isolation between Pedicularis groenlandica and P. attollens (Scrophulariaceae). Biologisches Zentralblatt 113:43–51.
95. Guzman B, Lledo MD, Vargas P. 2009. Adaptive radiation in Mediterranean Cistus (Cistaceae). PLOS One 4:e6362. doi: 10.1371/journal.pone.0006362.
96. Hansen T, Armbruster WS, Antonsen L. 2000. Comparative analysis of character displacement and spatial adaptations as illustrated by the evolution of Dalechampia blossoms. American Naturalist 156:S17–S34.
97. Hansen TF, Pe´labon C, Armbruster WS, Carlson ML. 2003a. Evolvability and genetic constraint in Dalechampia blossoms: components of variance and measures of evolvability. Journal of Evolutionary Biology 16:754–766.
98. Hansen TF, Armbruster WS, Carlson ML, Pe´labon C. 2003b. Evolvability and genetic constraint in Dalechampia blossoms: genetic correlations and conditional evolvability. Jornal of Experimental Zoology: Molecular and Developmental Evolution 296B: 23–39.
99. Hansen TF, Carter AJR, Pelabon C. 2006. On adaptive accuracy and precision in natural populations. American Naturalist 168: 168–181.
100. Hansen TF, Pelabon C, Armbruster WS. 2007. Comparing variational properties of homologous floral and vegetative characters in Dalechampia scandens: testing the Berg hypothesis. Evolutionary Biology 34:86–98.
101. Hapeman JR, Inoue K. 1997. Plant–pollinator interactions and floral radiation in Platanthera (Orchidaceae). In: Givnish TJ, Sytsma KJ, eds. Molecular evolution and adaptive radiation. Cambridge, UK: Cambridge University Press, 433–454.
102. Harmon LJ, Schulte JA, Larson A, Losos JB. 2003. Tempo and mode of evolutionary radiation in Iguanian lizards. Science 301: 961–964.
103. Heinrich B. 1976. Foraging specializations of individual bumblebees. Ecological Monographs 46:105–128.

104. Heinrich B, Raven PH. 1972. Energetics and pollination ecology. Science 176:597–602.

105. Hermann K, Klahre U, Moser M, Sheehan H, Mandel T, Kuhlemeier C. 2013. Tight genetic linkage of prezygotic barrier loci creates a multifunctional speciation island in Petunia. Current Biology 23: 873–877.

106. Herrera CM. 2001. Deconstructing a floral phenotype: do pollinators select for corolla integration in Lavandula latifolia? Journal of Evolutionary Biology 14:574–584.

107. Herrera CM, Cerda X, Garcia MB, Guitian J, Medrano M, Rey PJ, Sanchez-Lafuente AM. 2002. Floral integration, phenotypic covariance structure and pollinator variation in bumblebeepollinated Helleborus foetidus. Journal of Evolutionary Biology 15:108–121.

108. Hodges SA. 1997. Floral nectar spurs and diversification. International Journal of Plant Sciences 158:S81–S88.

109. Hodges SA, Arnold ML. 1995. Spurring plant diversification: are floral nectar spurs a key innovation? Proceedings of the Royal Society B Biological Sciences 262:343–348.

110. Huang S-Q, Shi X-Q. 2013. Floral isolation in Pedicularis: how do congeners with shared pollinators minimize reproductive interference? New Phytologist 199:858–865.

111. Hughes M, Gerber S, Wills MA. 2013. Clades reach highest morphological disparity early in their evolution. Proceedings of the National Academy of Sciences of the USA 110: 13875–13879.

112. Hunter JP. 1998. Key innovations and the ecology of macroevolution. Trends in Ecology and Evolution 13:31–36.

113. Igic B, Busch JW. 2013. Is self-fertilization an evolutionary dead end? New Phytologist 198:386–397.

114. Inoue K. 1983. Systematics of the genus Platanthera (Orchidaceae) in Japan and adjacent regions with special reference to pollination. Journal of the Faculty of Science University of Tokyo Section III Botany 13:285–374.

115. Irwin RE, Bronstein JL, Manson JS, Richardson L. 2010. Nectar robbing: ecological and evolutionary perspectives. Annual Review of Ecology, Evolution, and Systematics 41:271–292.

116. Johnson SD. 1997. Pollination ecotypes of Satyrium hallackii (Orchidaceae) in South Africa. Botanical Journal of the Linnean Society 123:225–235.

117. Johnson SD. 2006. Pollinator-driven speciation in plants. In: Harder LD, Barrett SCH, eds. Ecology and evolution of flowers. Oxford, UK: Oxford University Press, 295–310.

118. Johnson SD. 2010. The pollination niche and its role in the diversification and maintenance of the southern African flora. Philosophical Transactions of the Royal Society B—Biological Sciences 365: 499–516.

119. Johnson SD, Steiner KE. 1997. Long-tongued fly pollination and evolution of floral spur length in the Disa draconis complex (Orchidaceae). Evolution 51:45–53.

120. Johnson SD, Steiner KE. 2000. Generalization versus specialization in plant pollination systems. Trends in Ecology and Evolution 15: 140–143.

121. Johnson SD, Linder HP, Steiner KE. 1998. Phylogeny and radiation of pollination systems in Disa (Orchidaceae). American Journal of Botany 85:402–411.

122. Kalisz S, Vogler D, Fails B, Finer M, Shepard E, Herman T, Gonzales R. 1999. The mechanism of delayed selfing in Collinsia verna (Scrophulariaceae). American Journal of Botany 86:1239–1247.

123. Kay KM, Sargent RD. 2009. The role of animal pollination in plant speciation: integrating ecology, geography, and genetics. Annual Review of Ecology, Evolution, and Systematics 40: 637–656.

124. Kay KM, Reeves PA, Olmstead RG, Schemske DW. 2005. Rapid speciation and the evolution of hummingbird pollination in neotropical Costus subgenus Costus (Costaceae): evidence from nrDNA ITS and ETS sequences. American Journal of Botany 92: 1899–1910.

125. Klak C, Reeves G, Hedderson T. 2004. Unmatched tempo of evolution in Southern African semi-desert ice plants. Nature 427:63–65.

126. Kochmer JP, Handel SN. 1986. Constraints and competition in the evolution of flowering phenology. Ecological Monographs 56: 303–325.

127. Lambrecht SC. 2013. Floral water costs and size variation in the

highly selfing Leptosiphon biolor (Polemoniaceae). International Journal of Plant Sciences 174:74–84.

128. Lambrecht SC, Dawson TE. 2007. Correlated variation of floral and leaf traits along a moisture availability gradient. Oecologia 151: 574–583.

129. Lamont BB, He TH. 2012. Fire-adapted Gondwanan angiosperm floras evolved in the Cretaceous. BMC Evolutionary Biology 12: 223.

130. Lankinen A , Armbruster WS. 2007. Pollen competition reduces inbreeding depression in Collinsia heterophylla (Plantaginaceae). Journal of Evolutionary Biology 20:737–749.

131. Ley AC, Classen-Bockhoff R. 2009. Pollination syndromes in African Marantaceae. Annals of Botany 104:41–56.

132. Litman JR, Danforth BN, Eardley CD, Praz CJ. 2011. Why do leafcutter bees cut leaves? New insights into the early evolution of bees. Proceedings of the Royal Society B Biological Sciences 278: 3593–3600.

133. Lumbsch HT, Parnmen S, Rangsiruji A, Elix JA. 2010. Phenotypic disparity and adaptive radiation in the genus Cladia (Lecanorales, Ascomycota). Australian Systematic Botany 23:239–247.

134. Maad J, Armbruster WS, Fenster CB. 2013. Floral evolution in Campanula rotundifolia (Campanulaceae) along altitudinal gradients: patterns and possible selective mechanisms. Nordic Journal of Botany 31:361–371.

135. Macior LW. 1983. The pollination dynamics of sympatric species of Pedicularis (Scrophulariaceae). American Journal of Botany 70: 844–853.

136. Marten-Rodriguez S, Fenster CB, Agnarsson I, Skog LE, Zimmer EA. 2010. Evolutionary breakdown of pollination specialization in a Caribbean plant radiation. New Phytologist 188:403–417.

137. Mayfield MM,Waser NM, PriceMV. 2001. Exploring the 'most effective pollinator principle' with complex flowers: Bumblebees and Ipomopsis aggregata. Annals of Botany 88:591–596.

138. Mazer SJ, Paz H, Bell MD. 2004. Life history, floral development and mating system in Clarkia xantiana (Onagraceae): do floral and whole-plant rates of development evolve independently? American Journal of Botany 91:2041–2050.

139. McCall AC, Irwin RE. 2006. Florivory: the intersection of pollination and herbivory. Ecology Letters 9:1351–1365.
140. Merriam-Webster. 2013. Online dictionary. http://www. merriamwebster. com/dictionary (accessed 7 October 2013).
141. Miller RB. 1981. Hawkmoths and the geographic patterns of floral variation in Aquilegia caerulea. Evolution 35:763–774.
142. Moeller DA, Geber MA. 2005. Ecological context of the evolution of self-pollination in Clarkia xantiana: population size, plant communities, and reproductive assurance. Evolution 59: 786–799.
143. Morales M, Ackerman JD, Tremblay RL. 2010. Morphological flexibility across an environmental gradient in the epiphytic orchid, Tolumnia variegata: complicating patterns of fitness. Botanical Journal of the Linnean Society 163:431–446.
144. Moyle LC, Olson MS, Tiffin P. 2004. Patterns of reproductive isolation in three angiosperm genera. Evolution 58:1195–1208.
145. Muchhala N. 2007. Adaptive trade-off in floral morphology mediates specialization for flowers pollinated by bats and hummingbirds. American Naturalist 169:494–504.
146. Muchhala N, Potts MD. 2007. Character displacement among batpollinated flowers of the genus Burmeistera: analysis of mechanism, process and pattern. Proceedings of the Royal Society B Biological Sciences 274:2731–2737.
147. Muchhala N, Brown Z, Armbruster WS, Potts MD. 2010. Competition drives specialization in pollination systems through costs to male fitness. American Naturalist 176:732–743.
148. Mulcahy DL. 1979. Rise of the angiosperms - genecological factor. Science 206:20–23.
149. Murren CJ. 2012. The integrated phenotype. Integrative and Comparative Biology 52:64–76.
150. Nakazato T, Rieseberg LH, Wood TE. 2013. The genetic basis of speciation in the Giliopsis lineage of Ipomopsis (Polemoniaceae). Heredity 111:227–237.
151. Nattero J, Sersic AN, Cocucci AA. 2011. Geographic variation of floral traits in Nicotiana glauca: relationships with biotic and abiotic factors. Acta Oecologica 37:503–511.

152. Nosil P. 2012. Ecological speciation. Oxford, UK: Oxford University Press.

153. Ollerton J. 1996. Reconciling ecological processes with phylogenetic patterns: the apparent paradox of plant–pollinator systems. Journal of Ecology 84:767–769.

154. Ollerton J, Killick A, Lamborn E, Watts S, Whiston M. 2007. Multiple meanings and modes: on the many ways to be a generalist flower. Taxon 56:717–728.

155. Ordano M, Fornoni J, Boege K, Dominguez CA. 2008. The adaptive value of phenotypic floral integration. New Phytologist 179: 1183–1192.

156. Pannell JR, Labouche AM. 2013. The incidence and selection of multiple mating in plants. Philosophical Transactions of the Royal Society B—Biological Sciences 368:Article Number 20120051. doi:10.1098/rstb.2012.0051.

157. Pansarin ER, do Amaral MDE. 2009. Reproductive biology and pollination of southeastern Brazilian Stanhopea Frost ex Hook (Orchidaceae). Flora 204:238–249.

158. Parra-Tabla V, Vargas CF. 2004. Phenology and phenotypic natural selection on the flowering time of a deceit-pollinated tropical orchid, Myrmecophila christinae. Annals of Botany 94:243–250.

159. Pe´labon C, Armbruster WS, Hansen TF. 2011. Experimental evidence for the Berg hypothesis: vegetative traits are more sensitive than pollination traits to environmental variation. Functional Ecology 25:247–257.

160. Pe´labon C, Armbruster WS, Hansen TF, Bolstad GH, Pe´rez-Barrales R. 2012. Adaptive accuracy and adaptive landscapes. In: Svensson E, Calsbeek R, eds. The adaptive landscape in evolutionary biology. Oxford, UK: Oxford University Press, 150–168.

161. Perez F, Arroyo MTK, Medel R, Hershkovitz MA. 2006. Ancestral reconstruction of flower morphology and pollination systems in Schizanthus (Solanaceae). American Journal of Botany 93: 1029–1038.

162. Pe´rez-Barrales R, Arroyo J, Armbruster WS. 2007. Differences in pollinator faunas may generate geographic differences in

floral morphology and integration in Narcissus papyraceus (Amaryllidaceae). Oikos 116:1904–1918.

163. Pe´rez-Barrales R, Bolstad GH, Pe´labon C, Hansen TF, Armbruster WS. 2013. Pollinators and seed predators generate conflicting selection on Dalechampia blossoms. Oikos 122:1411–1428.

164. Peter CI, Johnson SD. 2014. A pollinator shift explains floral divergence in an orchid species complex in South Africa. Annals of Botany 113:277–288.

165. Pettengill JB, Moeller DA. 2012. Tempo and mode of mating system evolution between incipient Clarkia species. Evolution 66: 1210–1225.

166. Pohl NB, Van Wyk J, Campbell DR. 2011. Butterflies show flower colour preferences but not constancy in foraging at four plant species. Ecological Entomology 36:290–300.

167. Rabosky DL, Matute DR. 2013. Macroevolutionary speciation rates are decoupled from the evolution of intrinsic reproductive isolation in Drosophila and birds. Proceedings of the National Academy of Sciences of the USA 110:15354–15359.

168. Rabosky DL, McCune AR. 2010. Reinventing species selection with molecular phylogenies. Trends in Ecology & Evolution 25: 68–74.

169. Rabosky DL, Santini F, Eastman J, Smith SA, Sidlauskas B, Chang J, Alfaro ME. 2013. Rates of speciation and morphological evolution are correlated across the largest vertebrate radiation. Nature Communications 4:Article Number 1958. doi:10.1038/ncomms2958.

170. Raine NE, Chittka L. 2005. Comparison of flower constancy and foraging performance in three bumblebee species (Hymenoptera: Apidae: Bombus). Entomologia Generalis 28:81–89.

171. Raine NE, Pierson AS, Stone GN. 2007. Plant–pollinator interactions in a Mexican Acacia community. Arthropod–Plant Interactions 1: 101–117.

172. Ramsey J, Bradshaw HD, Schemske DW. 2003. Components of reproductive isolation between the monkey flowers Mimulus lewisii and M. cardinalis (Phrymaceae). Evolution 57:1520–1534.

173. Rathcke B, Lacey EP. 1985. Phenological patterns of terrestrial plants. Annual Review of Ecology and Systematics 16:179–214.

174. Raven PH, Axelrod DI. 1974. Angiosperm biogeography and past continental movements. Annals of the Missouri Botanical Garden 61: 539–673.

175. Rees M, Ellner SP. 2009. Integral projection models for populations in temporally varying environments. Ecological Monographs 79: 575–594.

176. Regal PJ. 1977. Ecology and evolution of flowering plant dominance. Science 196:622–629.

177. Robertson JL, Wyatt R. 1990. Evidence for pollination ecotypes in the yellow-fringed orchid, Platanthera ciliaris. Evolution 44: 121–133.

178. Roughgarden J. 1979. Theory of population genetics and evolutionary ecology: an introduction. New York: MacMillan.

179. Runions CJ, Geber MA. 2000. Evolution of the self-pollinating flower in Clarkia xantiana (Onagraceae). I. Size and development of floral organs. American Journal of Botany 87:1439–1451.

180. Sakai S, Kawakita A, Ooi K, Inoue T. 2013. Variation in the strength of association among pollination systems and floral traits: evolutionary changes in the floral traits of Bornean gingers (Zingiberaceae). American Journal of Botany 100:546–555.

181. Santos-Gally R, Perez-Barrales R, Simon VI, Arroyo J. 2013. The role of short-tongued insects in floral variation across the range of a style-dimorphic plant. Annals of Botany 111:317–328.

182. Sargent RD. 2004. Floral symmetry affects speciation rates in angiosperms. Proceedings of the Royal Society B Biological Sciences 271:603–608.

183. Schemske DW, Bradshaw HD. 1999. Pollinator preference and the evolution of floral traits in monkeyflowers (Mimulus). Proceedings of the National Academy of Sciences of the USA 96: 11910–11915.

184. Schemske DW, Horvitz CC. 1984. Variation among floral visitors in pollination ability—a precondition for mutualism specialization. Science 225:519–521.

185. Schemske DW, Horvitz CC. 1989. Temporal variation in selection on a floral character. Evolution 43:461–465.

186. Schiestl FP, Ayasse M. 2002. Do changes in floral odor cause

speciation in sexually deceptive orchids? Plant Systematics and Evolution 234:111–119.

187. Schiestl FP, Schlueter PM. 2009. Floral isolation, specialized pollination and pollinator behavior in orchids. Annual Review of Entomology 54:425–446.
188. Schluter D. 2000. The ecology of adaptive radiation. Oxford, UK: Oxford University Press.
189. Schmidt-Kittler N. 2002. Feeding specializations in rodents. Lethaea 82:141–152.
190. Scopece G, Musacchio A, Widmer A, Cozzolino S. 2007. Patterns of reproductive isolation in Mediterranean deceptive orchid. Evolution 61:2623–2642.
191. Siepielski AM, DiBattista JD, Carlson SM. 2009. It's about time: the temporal dynamics of phenotypic selection in the wild. Ecology Letters 12:1261–1276.
192. Silvertown J, Servaes C, Biss P, Macleod D. 2005. Reinforcement of reproductive isolation between adjacent populations in the Park Grass Experiment. Heredity 95:198–205.
193. Simpson GG. 1944. Tempo and mode in evolution. New York: Columbia University Press.
194. Simpson GG. 1953. The major features of evolution. New York: Columbia University Press.
195. Skogsmyr I, Lankinen A. 2002. Sexual selection: an evolutionary force in plants. Biological Reviews 77:537–562.
196. Sletvold N, Grindeland JM, Agren J. 2010. Pollinator-mediated selection on floral display, spur length and flowering phenology in the deceptive orchid Dactylorhiza lapponica. New Phytologist 188: 385–392.
197. Sletvold N, Grindeland JM, Agren J. 2013. Vegetation context influences the strength and targets of pollinator-mediated selection in a deceptive orchid. Ecology 94:1236–1242.
198. Smith SD. 2010. Using phylogenetics to detect pollinator-mediated floral evolution. New Phytologist 188:354–363.
199. Smith SD, Ane C, Baum DA. 2008. The role of pollinator shifts in the floral diversification of Iochroma (Solanaceae). Evolution 62: 793–806.

200. Stanley SM. 1975. A theory of evolution above the species level. Proceedings of the National Academy of Sciences of the USA 72: 646–650.
201. Stebbins GL. 1970. Adaptive radiation of reproductive characteristics in angiosperms. I. Pollination mechanisms. Annual Review of Ecology and Systematics 1:307–326.
202. Stebbins GL. 1974. Plant species. Evolution above the species level. Cambridge, MA: Harvard University Press, 62.
203. Stiles FG. 1975. Ecology, flowering phenology, and hummingbird pollination of some Costa Rican Heliconia species. Ecology 56: 285–301.
204. Stone GN, Willmer P, Rowe JA. 1998. Partitioning of pollinators during flowering in an African Acacia community. Ecology 79: 2808–2827.
205. Strauss SY, Irwin RE. 2004. Ecological and evolutionary consequences of multispecies plant–animal interactions. Annual Review of Ecology, Evolution, and Systematics 35: 435–466.
206. Strauss SY, Whittall JB. 2006. Non-pollinator agents of selection on floral traits. In: Harder KD, Barrett SCH, eds. Ecology and evolution of flowers. Oxford, UK: Oxford University Press, 120–138.
207. Svensson E, Calsbeek R. 2012. The adaptive landscape in evolutionary biology. Oxford, UK: Oxford University Press.
208. Temeles EJ, Koulouris CR, Sander SE, Kress WJ. 2009. Effect of flower shape and size on foraging performance and trade-offs in a tropical hummingbird. Ecology 90:1147–1161.
209. Temeles EJ, Rah YJ, Andicoechea J, Byanova KL, Giller GSJ, Stolk SB, Kress WJ. 2013. Pollinator-mediated selection in a specialized hummingbird–Heliconia system in the Eastern Caribbean. Journal of Evolutionary Biology 26:347–356.
210. Thomson JD, Wilson P. 2008. Explaining evolutionary shifts between bee and hummingbird pollination: convergence, divergence, and directionality. International Journal of Plant Science 169: 23–38.
211. Tripp EA, Manos PS. 2008. Is floral specialization an evolutionary dead-end? Pollination system transitions in Ruellia (Acanthaceae). Evolution 62:1712–1736.

212. Valiente-Banuet A, Molina-Freaner F, Torres A, Del Coro Arizmendi A, Casas A. 2004. Geographic differentiation in the pollination system of the columnar cactus Pachycereus pecten-aboriginum. American Journal of Botany 91:850–855.
213. van der Niet T, Johnson SD. 2012. Phylogenetic evidence for pollinator-driven diversification of angiosperms. Trends in Ecology and Evolution 27:353–361.
214. van der Niet T, Johnson SD, Linder HP. 2006. Macroevolutionary data suggest a role for reinforcement in pollination system shifts. Evolution 60:1596–1601.
215. Vereecken NJ, Streinzer M, Ayasse M, Spaethe J, Paulus HF, Stokl J, Cortis P, Schiestl FP. 2011. Integrating past and present studies on Ophrys pollination—a comment on Bradshaw et al. Botanical Journal of the Linnean Society 165:329–335.
216. Waser NM. 1986. Flower constancy—definition, cause, and measurement. American Naturalist 127:593–603.
217. Waser NM. 1998. Pollination, angiosperm speciation, and the nature of species boundaries. Oikos 82:198–201.
218. Waser NM. 2001. Pollinator behavior and plant speciation: looking beyond the 'ethological isolation' paradigm. In: Chittka L, Thomson JD, eds. Cognitive ecology of pollination. Cambridge, UK: Cambridge University Press, 318–335.
219. Waser NM, Chittka L, Price MV, Williams NM, Ollerton J. 1996. Generalization in pollination systems, and why it matters. Ecology 77: 1043–1060.
220. Wills MA, Briggs DEG, Fortey RA. 1994. Disparity as an evolutionary index: a comparison of Cambrian and Recent arthropods. Paleobiology 20:93–130.
221. Wilson P, Stine M. 1996. Floral constancy in bumble bees: handling efficiency or perceptual conditioning? Oecologia 106:493–499.
222. Wilson P, Thomson JD. 1996. How do flowers diverge? In: Lloyd DG, Barrett SCH, eds. Floral biology. NewYork: Chapman and Hall, 88–111.
223. Wilson P, Wolfe AD, Armbruster WS, Thomson JD. 2007. Constrained lability in floral evolution: counting convergent

origins of hummingbird pollination in Penstemon and Keckiella. New Phytologist 176:883–890.

224. Whalen MD. 1978. Reproductive character displacement and floral diversity in Solanum section Androceras. Systematic Botany 3: 77–86.
225. Wright S. 1951. The genetical structure of populations. Annals of Eugenics 15:323–354.
226. Yuan YW, Byers KJRP, Bradshaw HD. 2013. The genetic control of flower–pollinator specificity. Current Opinion in Plant Biology 16:422–428.
227. Zhao ZG, Huang SQ. 2013. Differentiation of floral traits associated with pollinator preference in a generalist-pollinated herb, Trollius ranunculoides (Ranunculaceae). International Journal of Plant Sciences 174:637–646.

Chapter 6

A Functional Role for the Colleters of Coffee Flowers

Juliana Lischka Sampaio Mayer, Sandra Maria Carmello-Guerreiro, and Paulo Mazzafera

Departamento de Biologia Vegetal, Instituto de Biologia, CP 6109, Universidade Estadual de Campinas, 13083-970 Campinas, SP, Brazil

ABSTRACT

Colleters are protuberances or trichomes that produce and release an exudate that overlays vegetative or reproductive buds. Colleters have a functional definition, as they are thought to protect young tissues against dehydration and pest attack. Decaffeinated coffee plants, named Decaffito®, have recently been obtained through chemical mutagenesis, and in addition to the absence of the alkaloid, the flowers of these plants open precociously. Decaffito mutants exhibit minimal production and secretion of the exudate by the colleters. We compared these mutants with normal coffee plants to infer the functional role of colleters and the secreted exudate covering flower buds. Decaffito mutants were obtained by sodium azide mutagenesis of Coffea arabica cv. Catuaí seeds. Wild-type plants were used as controls and are referred to as Catuaí. The flower colleters were analysed by scanning and transmission electron microscopy in addition to histochemical analysis. Histochemical analysis indicated the presence of heterogeneous exudate in the secretory cells of the colleters of both variants of coffee trees. Alkaloids were detected in Catuaí but not in Decaffito. Transmission electron microscopy revealed that the secretory cells in the Catuaí colleters possessed the normal and common characteristics found in secretory structures. In the secretory cells of the Decaffito colleters, it was not possible to identify any organelles or even the nucleus, but the cells had a darkened central cytoplasm, indicating that the secretion is produced in low amounts but not released. Our results offer a proof of concept of colleters in coffee, strongly indicating that the exudate covering the flower parts works as an adhesive to keep the petals together and the flower closed, which in part helps to avoid dehydration. Additionally, the exudate itself helps to prevent water loss from the epidermal cells of the petals.

INTRODUCTION

Colleters have been defined as secretory structures present in different organs of members of >60 angiosperm families, including Rubiaceae, Loganiaceae and Apocynaceae (Thomas and Dave 1990; Miguel et al. 2009). Structurally, they can appear as trichomes or as emergences that are formed from both epidermal and subepidermal tissues (Foster 1949; Appezzato-da-Gloria and Estelita 2000; Evert 2007).

Morphology, location and the chemical nature of the exudate are the criteria used to define the term colleter, but in fact the functional concept is the common link connecting most reports of this structure (Mayer et al. 2011). The term colleter originates from the Greek word 'colla', which means glue (Foster 1949). The nature of the sticky resinous or mucilaginous substance released by the colleters is diverse, and polysaccharides, proteins and lipids have been described as components (Miguel et al. 2006). These structures differentiate early, and their function seems to be to provide physical or chemical defences for the shoot apex and lateral buds against insect and pathogen attack. However, the occurrence of colleters is not limited to these plant parts, as they also occur in reproductive organs and seedlings (De-Paula and Oliveira 2007; Mayer et al. 2011). Other reports suggest that colleters and the exudate protect juvenile plant structures against dehydration because the exudate permeates and overlays the entire meristem and juvenile organs (Thomas and Dave 1989; Thomas et al. 1989; Klein et al. 2004). Appezzato-da-Gloria and Estelita (2000) argued that the exudate prevents water loss in hot tropical climates. In addition to dehydration protection, it has also been suggested that by covering the shoot apex meristem, the exudate may act as a physical barrier (Miguel et al. 2010). It is important to supplement morphological, structural and developmental data with information on the secretory product from histochemical tests or chemical analysis to construct an association between the definition of colleter and its functionality (Radford et al. 1974).

Although the alleged role of the colleter and its exudate is to protect plant parts against water loss and attack by insects and pathogens, only indirect evidence of this role has been obtained. Only a few reports have analysed the composition of the resinous material and have suggested such a role (Miguel et al. 2006 and references therein). According to Miguel et al. (2006), who demonstrated an in vitro fungicide property (spore germination) of the exudate from the colleters of Bathysa nicholsonii K. Schum. (Rubiaceae), the exudate protects the shoot apical meristem against pathogen attack. Another function attributed to the exudate from colleters is related to nutritional aspects regarding bacterial leaf nodule symbiosis with the Rubiaceae species (Horner and Lersten 1968; Lersten 1975).

The colleters in the stipules of Coffea arabica (Patel and Zaveri 1975) are classified as 'standard type' (Lersten 1974a, b). This type of colleter is formed by a secretory epidermis and the central axis of parenchyma cells, without vascular tissue. The origin of this type of colleter involves the protoderm and the ground meristem, as described for the colleters of Caryocar brasiliense Camb. (Paiva and Machado 2006a).

Under Brazilian climate conditions, coffee (C. arabica) floral buds start to differentiate from axillary buds in January at the leaf axils that pre-formed in August of the previous year (Majerowicz and Söndahl 2005). During the shorter days of April, the induction of the existing leaf buds to flower buds intensifies. Once they have developed into mature buds, they become dormant. Dormancy coincides with the start of the dry season, and as soon as the first rains of spring begin, flowering is triggered and anthesis occurs (Camargo and Camargo 2001). During development, coffee buds/flowers are covered by a viscous exudate of unknown composition. The functional role of this secretion has never been proved, but it is argued that it protects the dormant coffee floral bud from dehydration during the dry season, as has been suggested for many other plant species.

Recently, we used sodium azide to mutagenize coffee seeds, aiming to obtain decaffeinated coffee plants (P. Mazzafera, unpubl. res.; Borrell 2012). Among the ~33 000 seedlings analysed, seven were found not to contain caffeine (1,3,7-trimethylxanthine) due to a blockade in the methylation of theobromine (3,7-dimethylxanthine) mediated by caffeine synthase (Mazzafera et al. 1994). These plants were grown in the field for 2 years until they blossomed, at which point it was observed that, in addition to lacking caffeine, they exhibited precocious flower opening (see Fig. 1A). The term anthesis will not be used here because we understand that what happens in Decaffito is not a normal process. Although this process happens very early during bud development, the buds curiously also undergo a period of dormancy, i.e. they stop growing during the dry season and start to swell and increase in size with the first rains of spring. The flowers are smaller than normal flowers (Fig. 1B), but they produce viable pollen. By crossing these mutants with caffeinated plants, the descendants recover the normal caffeine content. Normal C. arabica plants have cleistogamy, and therefore self-pollination is high (Carvalho 1988), an advantage lost by the decaffeinated mutants as they are more prone to cross-pollination. Controlled crosses showed that every time the decaffeinated mutant was obtained, this early flower opening phenotype was displayed, suggesting a strong genetic link. These mutants were named Decaffito (Mazzafera et al. 2009).

Figure 1: (A) Flowers of Decaffito on a branch of a plant in the field and (B) open flowers of Catuaí and Decaffito.

Here, we report that the flowers of Decaffito open precociously because of the lack of exudate released by colleters. In normal flowers, the viscosity of the exudate seems to hold the petals together, acting as an adhesive, and does not allow them to open until they absorb water, swell and can then overcome the barrier imposed by the exudate. Furthermore, the exudate seems to protect against dehydration through the formation of a thick layer on the young flower buds, which have numerous stomata on the external petal surface. This information is the first direct evidence for a functional role of colleters and their exudate.

METHODS

Botanical Materials

Coffea arabica cv. Catuaí Vermelho and cv. Decaffito were used in this study. The plants were grown in the experimental field of the Department of Plant Biology of the State University of Campinas, Campinas-SP, Brazil. Catuaí is a commercial cultivar and contains ~1.2 % caffeine on a dry weight (DW) basis in its fruits (Guerreiro Filho and Mazzafera 2003) and 0.8 % DW in its leaves (Guerreiro Filho and Mazzafera 2000).

Decaffito mutants were obtained by treating Catuaí Vermelho seeds with sodium azide (0.003 or 0.01 % in 200 mM sodium phosphate pH 3, 48 h) and then germinating them in a sand bed. Approximately 33 000 plants were analysed for caffeine in the leaves using high-performance liquid chromatography, and seven were found to be almost devoid of caffeine. These plants were then transferred to field conditions (P. Mazzafera, unpubl. res.; Borrell 2012). Samples of the flower buds and flowers at different developmental stages were collected from these plants and used in our studies.

Light Microscopy

Samples were fixed under vacuum as described by Karnovsky (1965; modified by preparation in phosphate buffer pH 7.2) for 24 h and dehydrated in an ethanol series (10, 30, 50 and 70 %) and then in a tert-butyl alcohol (TBA) series (70, 85, 95 and 100 %) (Johansen 1940) for 48 h in each solution. The last dehydration in 100 % TBA was repeated three times. A three-fourths volume of solid Paraplast X-tra® (Fisher) was added to the samples in 100 % TBA, and the mixture was maintained at 58 °C. The Paraplast was changed three times, every 12 h. The samples were placed on moulds to solidify, and serial sections (5μm thick) were cut on a rotary microtome (Leica) and distended in heated plates at 48 °C. The Paraplast was removed by immersion of the slides in xylene, and the sections were subsequently rehydrated in absolute ethanol followed by distilled water. The sections were stained with safranin and astra-blue (Gerlach 1969) and mounted in Entellan®

synthetic resin (Merck). Photomicrographs were taken with an Olympus BX 51 photomicroscope equipped with an Olympus DP71 camera.

Histochemistry

Samples were fixed, dehydrated and embedded as described above. The chemical nature of the substances found in the secretory cells of the colleters and the exudate was determined using the following histochemical tests: periodic acid–Schiff's reaction for 1,2-glycol groups present in polysaccharides (McManus 1948); ruthenium red for acid polysaccharides and pectic substances (Johansen 1940); Wagner's reagent for alkaloids (Wagner et al. 1984); aniline blue black (Fisher 1968) to identify proteins; and Sudan black B (Pearse 1985) and Nile blue (Cain 1947) for neutral (stained pink) and acidic (stained blue) lipids to identify the aliphatic compounds. Standard control procedures were performed simultaneously.

Scanning Electron Microscopy

Samples were fixed as described by Karnovsky (1965; modified by preparation in phosphate buffer pH 7.2) for 24 h, dehydrated in a graded ethanol series and subjected to critical point drying with CO_2 (Horridge and Tamm 1969). The samples were attached to aluminium stubs and coated with gold (30–40 nm). Finally, the samples were examined under a LEO model VP 435 scanning electron microscope at 20 kV.

Transmission Electron Microscopy

Samples of bract with colleters were fixed by Karnovsky's method (Karnovsky 1965; modified by preparation in phosphate buffer pH 7.2), post-fixed in 1 % osmium tetroxide (0.1 M phosphate buffer pH 7.2) for 2 h, dehydrated by an acetone series and embedded in Araldite resin (Roland 1978; Silva and Machado 1999). Ultrathin sections were contrasted with uranyl acetate and lead citrate (Roland 1978) and examined under a Philips EM 100 transmission electron microscope at 60 kV.

RESULTS

Morphology of the Bud Flower

The Catuaí flower buds had a whitish, thick, viscous exudate covering the petals at different developmental stages before flower opening (Figs 2A, C and 3A). Although the exudate was not visible in fresh samples of Decaffito flowers (Figs 1A and 2B, D), analysis by scanning electron microscopy revealed its presence in minimal amounts (Figs 3B and 4G) when compared with Catuaí. The lack of exudate in Decaffito was observed from the beginning of the development of the reproductive meristem until the precocious opening

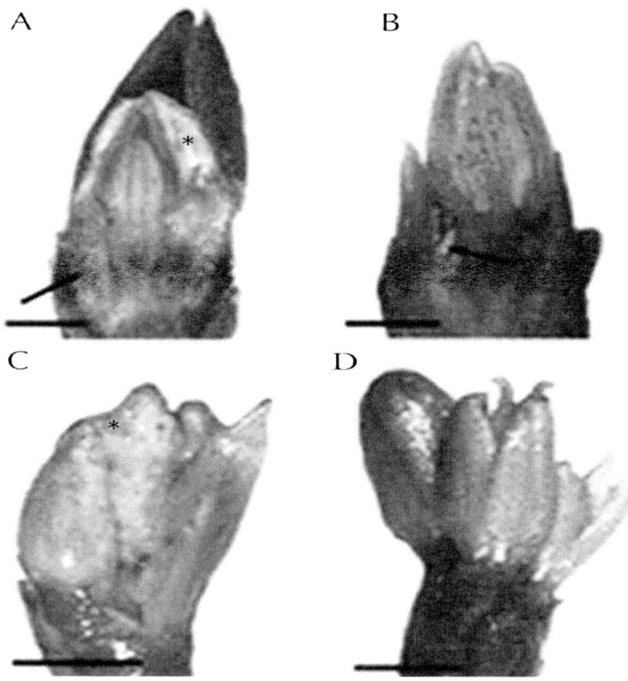

Figure 2: Flower buds of (A and C) Catuaí and (B and D) Decaffito. (A and B) Note the colleters (arrows) and the exudate covering the petals (*). Scale bars: A and B = 1 mm; C and D = 2 mm.

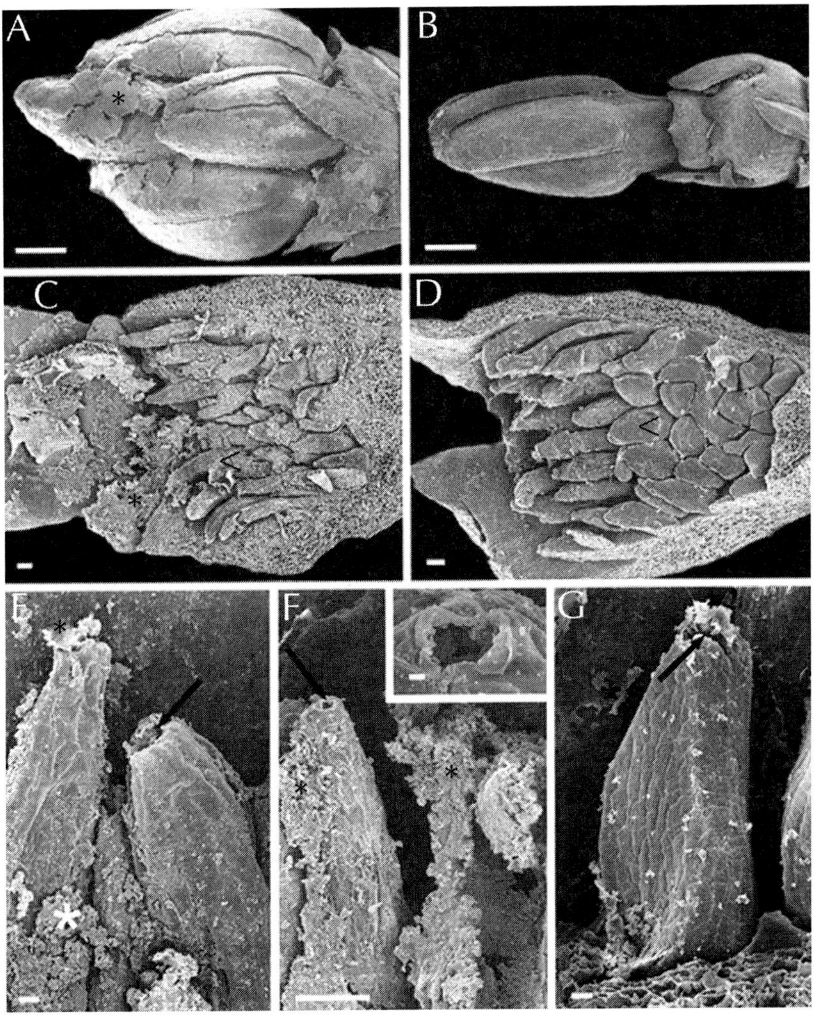

Figure 3: Scanning electron micrographs of flower buds and colleters. (A) Overview of the flower bud of Catuaí: observe the exudate covering the petals; (B) overview of the flower bud of Decaffito; (C and D) overview of colleters (arrowhead) in the bract; (C) Catuaí; (D) Decaffito; (E–G) details of colleters: note the cuticular rupture in the apical portion of the colleter (arrows), exposing the secretory cells and releasing the exudate (*); (E and F) Catuaí: note the apical portion of the colleter in detail; (G) Decaffito. Scale bars: A and B = 1 mm; C, D and F = 100 μm; E and G = 20 μm; F (inset) = 3 μm.

Figure 4: Scanning electron micrographs of the flowers of Decaffito. (A–F) Different stages of the flower bud; note the scarce presence of exudate; (D–F) flower in precocious opening; (G) note the scarce exudate (*) covering the petals; (H) overview of a flower bud, view of the colleter position (arrows) in the adaxial side of the bract; (I) details of colleters. Scale bars: A–F = 1 μm; G and I = 100 μm; H = 1 mm.

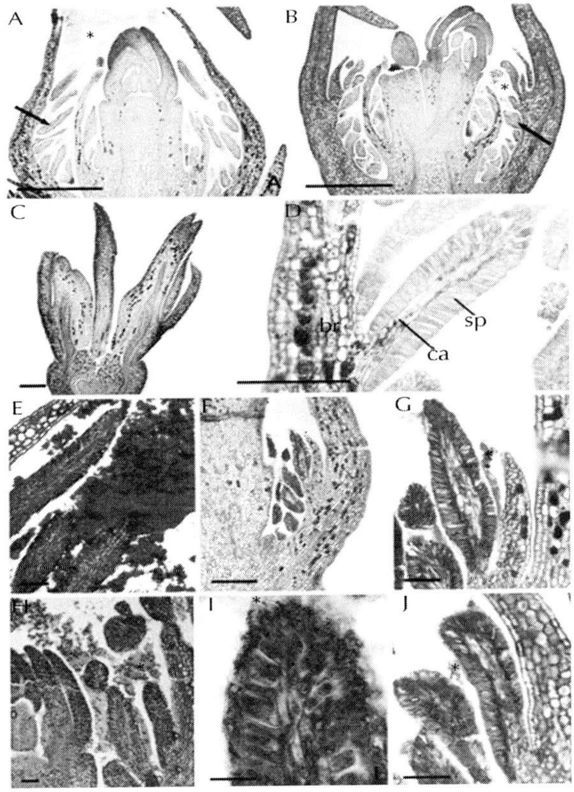

Figure 5: Longitudinal sections of flower buds. (A, E, H and I) Catuaí; (B–D, F, G and J) Decaffito. (A and B) Flower bud at the beginning of development; note the extracellular exudate (*) and colleter (arrows); (C) flower in precocious opening, without complete development of the floral organs; (D) details of the colleter with the secretory palisade epidermis (sp) and the central axis formed by non-secretory parenchyma cells (ca) in the adaxial side of the bract (br); (E–J) histochemical characterization of the exudates; (E–G) periodic acid–Schiff reaction; (H–J) ruthenium red. Scale bars: A and B = 500 μm; C and F = 200 μm; D = 100 μm; E, G, H and J = 50 μm; I = 20 μm.

The structures responsible for the secretion of the exudates are the colleters, which are positioned on the adaxial side of the bracts adjacent to the flower buds, as seen for Catuaí (Figs 2A, 3C and 5A, F) and Decaffito (Figs 2B, 3D, 4H, I and 5B–D). The colleters are long, have a short peduncle (Figs 3E–G and 4I) and are formed by a secretory palisade-like epidermis and an axis of non-secretory parenchyma central cells (Fig. 5D). The secretory phase of colleters begins at the induction of the reproductive meristem and remains active during the development of the flower bud.

HISTOCHEMISTRY

Histochemical tests revealed the complex and heterogeneous chemical nature of the exudate detected on the surface of the flower buds and inside the secretory cells of the colleters. The exudate is composed of polysaccharides (Fig. 5E–G), pectic substances (Fig. 5H–J), alkaloids (Fig. 6A and B), proteins (Fig. 6D–F) and lipophilic substances (Fig. 6G–L). The release of the exudate is abundant in Catuaí (Figs 5E, H, I and 6A, B, D, E, G, H, K, L) and scarce in Decaffito (Figs 5F, G, J and 6F, I, L). As expected, alkaloids were not detected in the exudate or inside the secretory cells of the colleters of Decaffito (Fig. 6C).

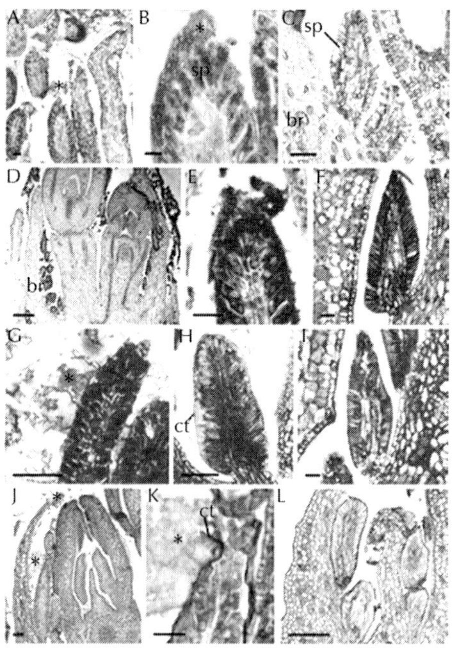

Figure 6: Histochemical characterization of the exudate of colleters. (A, B, D, E, G, H, J and K) Catuaí; (C, F, I and L) Decaffito. (A–C) Wagner's reagent: note

the secretory palisade epidermis (sp) and the extracellular exudate (*); (D–F) aniline blue black; (G–I) Nile blue; note in (H) the cuticle (ct) displacement and subcuticular space; (J–L) Sudan black B; note in (K) the cuticular rupture in the apical portion of the colleter, releasing the exudate (*). Scale bars: A, E, F, I and K = 20 μm; B = 10 μm; C, G, H and J = 50 μm; D = 200 μm; L = 100 μm.

The exudate released by the secretory epidermal cells between the cuticular layer and the wall layers below led to the formation of large subcuticular spaces in which the exudate accumulated (Fig. 6H). This accumulation creates pressure under the cuticle, causing it to rupture and release the exudate

Ultrastructure of Colleters

Catuaí

The secretory cells of the colleters of this coffee cultivar have a dense cytoplasm, an evident nucleus, small vacuoles, vesicles containing lipid-like substances, several dictyosomes, mitochondria and rough endoplasmic reticulum (Fig. 7A–E). The plastid matrix is granular with starch grains (clear bodies), classified as amyloplast (Fig. 7B and E). Some small vacuoles have an internal membrane system (Fig. 7C). The exudate is clearly heterogeneous (Fig. 7A, D and G), and it occupies intercellular spaces and the large subcuticular spaces formed by the separation of the cell wall from the cuticle. The non-secretory parenchyma cells of the central axis have a low-density content, and phenols are present (Fig. 7F). In the secretory cells near the senescent phase, the small vacuoles fuse, forming a large vacuole that occupies most of the protoplast interior (Fig. 7G). The cytoplasm becomes restricted to the periphery of the cell, and the rough endoplasmic reticulum appears to be parallel to the plasma membrane (Fig. 7H)

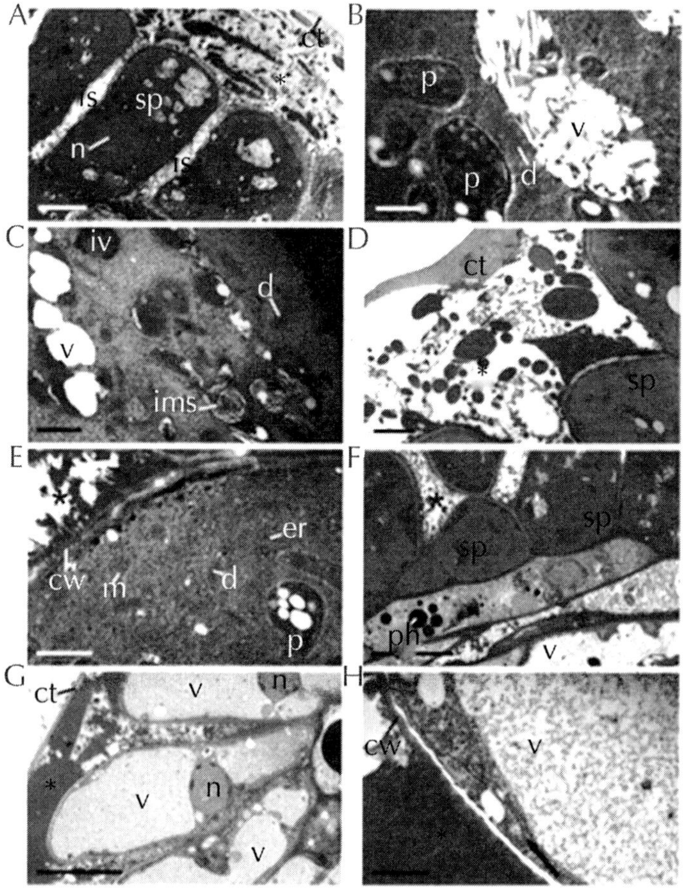

Figure 7: Transmission electron micrographs of Catuaí colleters. (A–F) Colleters in the secretory stage; (G and H) colleters close to senescence. (A) Secretory palisade epidermis (sp): note the dense cytoplasm, large nucleus (n), cuticle (ct) displacement and subcuticular space with exudate (*) and intercellular space (is); (B, C and E) details showing small vacuoles (v), vacuoles with an inner membrane system (ims), dictyosomes (d), plastids (p), lipid vesicles (lv), mitochondria (m) and rough endoplasmic reticulum (er); note in (D) the wide subcuticular space with heterogeneous exudate; note in (F) that the non-secretory parenchyma cell axis accumulates phenolic compounds; (G) secretory palisade epidermis close to senescence: note that the vacuoles increase in size and fuse; (H) the cytoplasm of these secretory cells consists of a thin peripheral layer close to the cell wall (cw): note the endoplasmic

reticulum parallel to the plasma membrane (arrows). Scale bars: A = 5 μm; B = 500 nm; C, E and H = 1 μm; D = 2 μm, F and G = 10 μm.

Decaffito

The secretory cells of the colleters of Decaffito show marked changes compared with Catuaí as shown by the ultrastructural analysis (Fig. 8A–E). Little exudate is produced, and it seems that it is either not released or when small amounts are released it is still enough to lead to the formation of subcuticular and intercellular spaces (Fig. 8A–C and E). Inside the cell, it is not possible to identify any organelles or even the nucleus, with only a darkened central cytoplasm (Fig. 8A and C–E) and a network of translucent tubule-like structures (Fig. 8C–E) visible. The non-secretory parenchyma cells of the central axis have characteristics similar to those of the Catuaí cell, with low-density content, large vacuoles and a detectable presence of the nucleus and plastids (Fig. 8F).

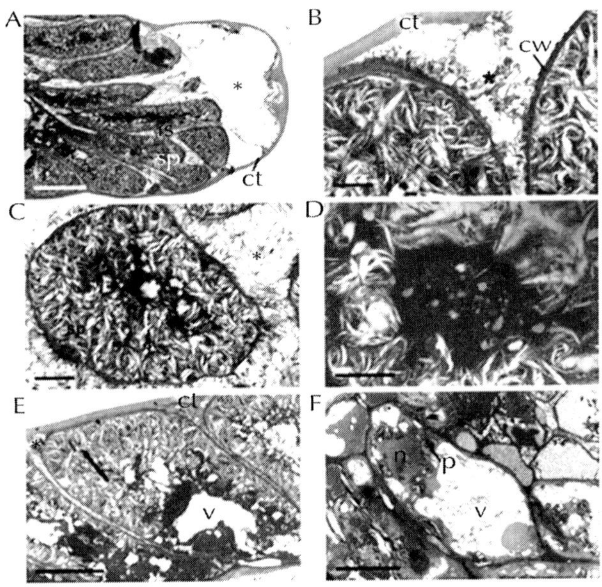

Figure 8: Transmission electron micrographs of Decaffito colleters. (A) Secretory palisade epidermis (sp) in the secretory stage; note the cuticle (ct)

displacement and subcuticular space with exudate (*) and intercellular space (is); (B) details of the cell wall (cw) of secretory cells and exudate; (C and D) cross-section of the secretory cell: note the completely disorganized cytoplasm; (D) details of the center of the secretory cell: note that no intact organelles were observed; only cytoplasm darkening was observed; (E) longitudinal section of the secretory cell: note the exudate inside (arrows) and outside (*) the cell and vacuoles (v); note in (F) the normal non-secretory parenchyma cell axis for the nucleus (n), plastids (p) and large vacuole (v). Scale bars: A and F = 10 μm; B and D = 1 μm; C = 2 μm; E = 5 μm.

DISCUSSION

Stipules protect vegetative buds and leaf primordia (Lubbock 1890; Paiva and Machado 2006a, b), and bracts protect reproductive structures (Bell 2008). Thus, the presence of colleters in the stipules (Patel and Zaveri 1975) and bracts of C. arabica reinforces the idea that they may have a protective role. Colleters occur exclusively on the adaxial face of bracts (Lersten 1974a) and are positioned above and adjacent to the reproductive meristem. The beginning of the secretory phase occurs prior to the development of flower organs, and as the flower buds start to develop, they are already covered by the exudate released from the colleters.

Both coffee plants in this study were grown in Campinas (SP, Brazil), where it is well established that coffee floral buds start to differentiate from axillary buds in January at leaf axils that pre-formed in August of the previous year. The formation of flower primordia occurs in March/April (Majerowicz and Söndahl 2005). Starting in May, the temperature and rain precipitation decline and remain low compared with other months until August/September. During this winter period, the vegetative growth of the flower buds ceases, and they remain dormant (Majerowicz and Söndahl 2005). The same climatic conditions apply to Catuaí and Decaffito, although the flower buds of the latter remain dormant, and no growth is observed. Because Decaffito does not have the presumed protection of the colleter exudate, the flower buds are exposed to low air humidity and may dehydrate. We suggest here that the colleter exudate in coffee also functions to keep the petals united, acting as an adhesive, by 'sealing' the bud. Once closed, the flowers may be partially protected from dehydration. In Decaffito, the petals are freed due to the lack of exudate. They also lose water from the external surface due to the presence of many stomata (see Fig. 4G), which may cause differential tension between the internal and external cell surfaces, forcing the petals to curve and open precociously (see Fig. 1A) before flower development is complete and when the flowers are still dormant. At the beginning of the rainy season, in September/October, the dormancy is broken and anthesis occurs after 10–12 days of the first rain (Majerowicz and Söndahl 2005). This is observed for both Catuaí and Decaffito, whose flowers are smaller than those of the former, most likely as a consequence of dehydration stress during the dry season (Fig. 1B). Thus, considering that Catuaí and Decaffito flowers differ regarding the presence of exudate, the Decaffito flowers provide functional proof of the role of colleters in protecting coffee flowers from dehydration and controlling their opening.

The histochemical evaluation showed that the difference in composition of the exudate of Catuaí and Decaffito was the absence of alkaloids in the latter, as would be expected given that Decaffito was selected for low caffeine content after sodium azide mutagenesis. The marked detection of caffeine in Catuaí is in agreement with chemical analysis, which showed that the caffeine content in the coffee flowers is among the highest of the different parts of the coffee tree (Hamidi and Wanner 1964; Raju and Gopal 1979).

The histochemical tests also revealed that the exudate composition was highly heterogeneous and complex. Polysaccharides, pectic substances, alkaloids, proteins and lipophilic substances were detected. Complex polysaccharide polymers of high molecular mass seem to play a role as an adhesive to aid in seed dispersion by fixing them to animals and by helping carnivorous plants to capture insects or to lubricate the root apex and facilitate interaction with microorganisms (Fahn 1988 and references therein). Additionally, the hydrophilic characteristics of these polymers seem to assist in maintaining appropriate humidity levels in the meristem and developing organs during dry periods, when soil water content and air humidity are low and temperatures are high (Kronestedt-Robards and Robards 1991; Paiva 2009). On the other hand, the colleter exudate is insoluble in water, which is most likely related to the lipid-like substances, produced to prevent water loss (Thomas and Dave 1989). The Catuaí exudate showed intense colouration for lipids. Kronestedt-Robards and Robards (1991) suggested that these substances could also inhibit the development of pathogenic microorganisms. Similarly, Paiva and Machado (2006a) suggested that proteins found in the exudate of colleters of C. brasiliense may have an anti-pathogenic function because they found enzyme activities related to chitinases and β-1,3-glucanases, which are usually related to protection against pathogens (Goy et al. 1992; Giannakis et al. 1998). The presence of protein in the exudate of colleters has also been related to the protection of meristems (Klein et al. 2004; Gonzalez and Tarragó 2009) but without a defined function. Although controversial (Guerreiro Filho and Mazzafera 2000, 2003), caffeine has been suggested to protect plants against insect attack (Nathanson 1984). However, it is noteworthy that at any time point, the Decaffito flowers were not observed as more likely to be under attack by insects or microorganisms than Catuaí or any other known C. arabica cultivar or coffee species containing caffeine (P. Mazzafera, pers. observ.), which suggests that, at least in coffee, the caffeine in the exudate from colleters does not have a function related to pest or pathogen protection.

The exudate covering the Catuaí flowers is mainly composed of polysaccharides and pectic compounds, which in turn seems to explain the presence of numerous dictyosomes in the secretory cells. Proteins were also densely stained. The secretion of protein–carbohydrate mucilage indicates participation of the Golgi complex and amyloplasts as well as the rough endoplasmic reticulum in the release process (Fahn 1988; Evert 2007). Amyloplasts are abundant in nectariferous tissue (Fahn 1988). They can act as organelles for the storage of substances necessary for the synthesis of the polysaccharide component of the nectar (Rachmilevitz and Fahn 1973; Nepi et al. 1996).

Epidermal cells of the colleters of Catuaí showed normal and common characteristics of secretory structures, with an evident nucleus, dense cytoplasm, various dictyosomes and mitochondria. However, the secretory cells of Decaffito did not show any distinguishable organelles, not even the nucleus, but only a darkened cytoplasm. In these cells, the exudate is produced in lower amounts than in Catuaí, and it is not secreted.

The exudate produced in Catuaí and Decaffito accumulates in subcuticular and intercellular spaces. Paiva and Machado (2006b) and Appezzato-da-Gloria and Estelita (2000) argued that such subcuticular spaces are formed by dissolution of the middle lamella due to enzyme activities along the anticlinal walls of the epithelial cells. Such processes increase the surface area from which the exudate is released as well as the space in which it can accumulate. Rupture of the cuticle by an increase in pressure caused by exudate accumulation in the subcuticular space has been observed in the colleters of other species of Rubiaceae (Thomas and Dave 1990), Caryocaraceae (Paiva and Machado 2006a) and Apocynaceae (Thomas and Dave 1989), which strongly suggests an absence of pores in the cuticle to facilitate exudate release.

Like any other secretory structure, colleters senesce after a secretory phase in which marked anatomical and ultrastructural alterations occur (Dickinson 2000). In the colleters of Catuaí, the main alteration observed with senescence was the state of the cytoplasm, from dense to less dense, and an enlargement of the vacuole. During the senescing phase of the colleters of B. nicholsonii, the secretory cells showed a disorganized system of endomembranes, and it was not possible to distinguish organelles, suggesting programmed cell death (Miguel et al. 2010). We could not visualize or distinguish any structural organization inside the secretory cells of the colleters of Decaffito in any phase. We speculate that such an occurrence is most likely related to a precocious programmed cell death process.

CONCLUSIONS

The Decaffito plants have very low caffeine content in all tissues, and this characteristic is profoundly associated with precocious flower opening (Borrell 2012; P. Mazzafera, unpubl. res.). Similar to natural mutants of C. arabica (Silvarolla et al. 2004), Decaffito plants accumulate theobromine, indicating a metabolic blockade of the last step of caffeine biosynthesis (Mazzafera et al. 2009). Although it is still not clear what controls caffeine biosynthesis in Decaffito coffee mutants, the associated and undesirable precocious flower opening characteristic provides the first functional proof of the role of colleters and their exudate in protecting flowers against exposure to dry atmospheres and acting as an adhesive to keep the petals united until anthesis. Additionally, although not reported here, we observed a lack of exudate on leaf buds, suggesting that the mutation in Decaffito might also affect the production and release of exudate in the colleters in different plant organs.

REFERENCES

1. Appezzato-da-Gloria B, Estelita MEM. Development, structure and distribution of colleters in Mandevilla illustris and M. velutina (Apocynaceae). Revista Brasileira de Botânica 2000;23:113-120.
2. Bell AD. Plant form: an illustrated guide to flowering plant morphology. 2nd edn. Portland: Timber Press; 2008.
3. Borrell B. Plant biotechnology: make it a decaf. Nature 2012;483:264-266.
4. Cain AJ. The use of Nile Blue in the examination of lipids. Quarterly Journal of Microscopical Science 1947;88:383-392.
5. Camargo AP, Camargo MBP. Definição e esquematização das fases fenológicas do cafeeiro arábica nas condições tropicais do Brasil. Bragantia 2001;60:65-68.
6. Carvalho A. Principles and practices of coffee plant breeding for productivity and quality factors: Coffea arabica. In: Clarke RJ, Macrae R, editors. Coffee: agronomy. London: Elsevier Applied Science; 1988. p. 129-165.
7. De-Paula OC, Oliveira DMT. Ocorrência de coléteres em embriões de três espécies de Chamaecrista Moench (Fabaceae: Caesalpinioideae). Revista Brasileira de Biociências 2007;5:348-350.
8. Dickinson WC. Integrative plant anatomy. San Diego: Harcourt Academic Press; 2000.
9. Evert RF. Esau's plant anatomy: meristems, cells, and tissues of the plant body, their structure, function and development. 2nd edn. Hoboken, NJ: John Wiley & Sons; 2007.
10. Fahn A. Secretory tissues in vascular plants. New Phytologist 1988;108:229-257.
11. Fisher DB. Protein staining of ribboned epon sections for light microscopy. Histochemie 1968;16:92-96.
12. Foster AS. Practical plant anatomy. 2nd edn. New York: D. Von Nostrand; 1949.
13. Gerlach G. Botanische Mikrotechnik, eine linfuhrung. Stuttgart: Georg Thieme; 1969.
14. Giannakis C, Bucheli CS, Skene KGM, Robinson SP, Scott NS.

Chitinase and β-1,3-glucanase in grapevine leaves: a possible defence against powdery mildew infection. Australian Journal of Grape and Wine Research 1998;4:14-22.

15. Gonzalez AM, Tarragó JR. Anatomical structure and secretion compounds of colleters in nine Ilex species (Aquifoliaceae) from southern South America. Botanical Journal of the Linnean Society 2009;160:197-210.
16. Goy PA, Felix G, Métraux JP, Meins F Jr.. Resistance to disease in the hybrid Nicotiana glutinosa× Nicotiana debneyi is associated with high constitutive levels of β-1,3-glucanase, chitinase, peroxidase and polyphenoloxidase. Physiological and Molecular Plant Pathology 1992;41:11-21.
17. Guerreiro Filho O, Mazzafera P. Caffeine does not protect coffee against the leaf miner Perileucoptera coffeella. Journal of Chemical Ecology 2000;26:1447-1464.
18. Guerreiro Filho O, Mazzafera P. Caffeine and resistance of coffee to the berry borer Hypothenemus hampei (Coleoptera: Scolytidae). Journal of Agricultural and Food Chemistry 2003;51:6987-6991.
19. Hamidi A, Wanner H. The distribution pattern of chlorogenic acid and caffeine in Coffea arabica. Planta 1964;61:90-96.
20. Horner HT, Lersten NR. Development, structure and function of secretory trichomes in Psychotria bacteriophila (Rubiaceae). American Journal of Botany 1968;55:1089-1099.
21. Horridge GA, Tamm SL. Critical point drying for scanning electron microscopy study of ciliary motion. Science 1969;163:817-818.
22. Johansen DA. Plant microtechnique. New York: McGraw Hill Book Company; 1940.
23. Karnovsky MJ. A formaldehyde-glutaraldehyde fixative of high osmolality for use in electron microscopy. The Journal of Cell Biology 1965;27:137-138.
24. Klein DE, Gomes VM, Silva-Neto SJ, Cunha M. The structure of colleters in several species of Simira (Rubiaceae). Annals of Botany 2004;94:733-740.
25. Kronestedt-Robards E, Robards A. Exocytosis in gland cells. In: Hawea C, Coleman J, Evans D, editors. Endocytosis, exocytosis and vesicle traffic in plants. Cambridge: Cambridge University Press; 1991. p. 199-232.

26. Lersten NR. Morphology and distribution of colleters and crystals in relation to the taxonomy and bacterial leaf nodule symbiosis of Psychotria (Rubiaceae). American Journal of Botany 1974;61:973-981.
27. Lersten NR. Colleter morphology in Pavetta, Neorosea and Tricalysia (Rubiaceae) and its relationship to the bacterial leaf nodule symbiosis. Botanical Journal of the Linnaean Society 1974;69:125-136.
28. Lersten NR. Colleter types in Rubiaceae, especially in relation to the bacterial leaf nodule symbiosis. Botanical Journal of the Linnaean Society 1975;71:311-319.
29. Lubbock J. On stipules, their form and function. Journal of the Linnean Society 1890;28:217-243.
30. Majerowicz N, Söndahl MR. Induction and differentiation of reproductive buds in Coffea arabica L. Brazilian Journal of Plant Physiology 2005;17:247-254.
31. Mayer JLS, Cardoso-Gustavson P, Appezzato-da-Glória B. Colleters in monocots: new record for Orchidaceae. Flora 2011;206:185-190.
32. Mazzafera P, Wingsle G, Olsson O, Sandberg G. S-adenosyl-L-methionine:theobromine 1-N-methyltransferase, an enzyme catalyzing the synthesis of caffeine in coffee. Phytochemistry 1994;37:1577-1584.
33. Mazzafera P, Baumann TW, Shimizu MM, Silvarolla MB. Decaf and the steeplechase towards Decaffito – the coffee from caffeine-free Arabica plants. Tropical Plant Biology 2009;2:63-76.
34. McManus JFA. Histological and histochemical uses of periodic acid. Stain Technology 1948;23:99-108.
35. Miguel EC, Gomes VM, Oliveira MA, Cunha MD. Colleters in Bathysa nicholsonii K. Schum. (Rubiaceae): ultrastructure, secretion protein composition and antifungal activity. Plant Biology 2006;8:715-722.
36. Miguel EC, Moraes DG, Cunha MD. Stipular colleters in Psychotria nuda (Cham. & Schltdl.) Wawra (Rubiaceae): micromorphology, anatomy and crystals microanalysis. Acta Botânica Brasílica 2009;23:1034-1039.
37. Miguel EC, Klein DE, Oliveira MA, Cunha M. Ultrastructure

of secretory and senescence phase in colleters of Bathysa gymnocarpa and B. stipulata (Rubiaceae). Revista Brasileira de Botânica 2010;33:425-436.

38. Nathanson JA. Caffeine and related methylxanthines: possible naturally occurring pesticides. Science 1984;226:184-187.

39. Nepi M, Ciampolini F, Pacini E. Development and ultrastructure of Cucurbita pepo nectaries of male flowers. Annals of Botany 1996;78:95-104.

40. Paiva EAS. Occurrence, structure and functional aspects of the colleters of Copaifera langsdorffii Desf. (Fabaceae, Caesalpinioideae). Comptes Rendus Biologies 2009;332:1078-1084.

41. Paiva EAS, Machado SR. Colleters in Caryocar brasiliense (Caryocaraceae) ontogenesis, ultrastructure and secretion. Brazilian Journal of Biology 2006;66:301-308.

42. Paiva EAS, Machado SR. Ontogenesis, structure and ultrastructure of Hymenaea stigonocarpa (Fabaceae: Caesalpinioideae) colleters. Revista de Biología Tropical 2006;54:943-950.

43. Patel JD, Zaveri M. Development of leaf and stipular glands in Coffea arabica. Flora 1975;164:11-18.

44. Pearse AGE. Histochemistry: theoretical and applied. 4th edn. Vol II. Edinburgh: Churchill Livingstone; 1985.

45. Rachmilevitz T, Fahn A. Ultrastructure of nectaries of Vinca rosea L., Vinca major L. and Citrus sinensis Osbeck cv. Valencia and its relation to the mechanism of nectar secretion. Annals of Botany 1973;37:1-9.

46. Radford AE, Dickison WC, Massey JR, Bell CR. Vascular plant systematics. New York: Harper & Row; 1974.

47. Raju KI, Gopal NH. Distribution of caffeine in arabica and robusta coffee plants. Journal of Coffee Research 1979;9:83-90.

48. Roland JC. General preparations and staining of thin sections. In: Hall JL, editor. Electron microscopy and cytochemistry of plant cells. New York: Elsevier; 1978. p. 1-62.

49. Silva EMJ, Machado SR. Ultrastructure and cytochemistry of pearl gland in Piper regnellii (Miq.) C. DC. – Piperaceae. Nordic Journal of Botany 1999;19:623-634.

50. Silvarolla MB, Mazzafera P, Fazuoli LC. A naturally decaffeinated arabica coffee. Nature 2004;429:826.

51. Thomas V, Dave Y. Histochemistry and senescence of colleters of Allamanda cathartica (Apocynaceae). Annals of Botany 1989;64:201-203.

52. Thomas V, Dave Y. Structure and necrosis of stipular colleters in Mitragyna parvifolia (Rubiaceae). Belgian Journal of Botany 1990;123:67-72.

53. Thomas V, Dave Y, Menon ARS. Anatomy and histochemistry of colleters in Roupelia grata (Apocynaceae). Nordic Journal of Botany 1989;8:493-496.

54. Wagner H, Bladt S, Zgainski EM. Plant drug analysis – a thin layer chromatography. Atlas, NY: Springer; 1984.

Chapter 7

Structural Organization of the Gynoecium and Pollen Tube Path in Himalayan Sea Buckthorn, Hippophae Rhamnoides (Elaeagnaceae)

Yash Mangla[1], Rajesh Tandon[1],
Shailendra Goel[1] and S. N. Raina[2]

[1]Department of Botany, University of Delhi, Delhi 110 007, India

[2]Amity Institute of Biotechnology, Amity University, Sector 125, Noida 210 303, Uttar Pradesh, India

ABSTRACT

Closure of carpels or angiospermy, a key developmental innovation, has been accomplished through different ontogenic routes among the flowering plants. The mechanism of angiospermy produces structural novelties in the gynoecium, which in turn affects the progamic phase. In this paper, we present the structural details of the gynoecium and functional attributes of the progamic phase of Hippophae rhamnoides, a dioecious species of Elaeagnaceae. The gynoecium is unicarpellate, and the carpel is dorsiventrally symmetric and conduplicate. The pollen tube path comprises a prominent, ventrally localized dry and non-papillate stigma, a pseudostyle and a dorsally protruded superior ovary. The pollen tube path in the stigmatic region is subdermal, and from the pseudostyle onwards, it resides over the epidermis of conduplicated margins. The epidermal cells along this region are secretory but produce sparse extracellular matrix. The tube approaches the solitary ovule through a tiny conduit in the carpel, the ventral pore. The duration of the entire progamic phase is ~72 h. The observed mean pollen tube length from stigma to ovule was 908.13 ± 180 μm and the mean tube growth rate was 18.75 μm h−1. The study demonstrates that sea buckthorn, a core eudicot, has a simple gynoecium with a pollen tube pathway that incorporates elements of both completely externalized and internalized compitum.

INTRODUCTION

A conspecific pollen–pistil interaction encompassing pollen germination on a stigma, pollen tube growth and guidance through the style followed by successful fertilization constitutes the progamic phase in *angiosperms* (Heslop-Harrison 1975a, b; Linskens 1986). The functional attributes of the phase, such as duration, male gametophyte competition and regulation of the number of pollen tubes (selection/ screening), are controlled by structural modifications that have accompanied the closure of carpels (*angiospermy*) in flowering plants (Bell 1995; Herrero and Hormaza 1996; Sage et al. 2009; Williams et al. 2010). Among these modifications, specialization of an elaborate receptive surface into a terminal stigma and the internalized pollen tube transmitting tract (*internal compitum*) are two of the conspicuous developmental events (Endress and Igersheim 2000; Williams 2009; Endress 2011). These processes have apparently followed different evolutionary lines (Endress 1982; Scutt et al. 2006) and may significantly reflect the habitat conditions or the pollination syndromes (biotic and/ or abiotic) which proved effective in maximizing fitness in a species

(Wang et al. 2011). The pollen tube path in a majority of plants has evolved from an *extragynoecial compitum* to an internalized one through syncarpy, which has offered greater advantages by providing favourable conditions and the necessary nutrition for a prolonged, lengthier and even distribution of tube growth inside the pistil (Labarca and Loewus 1973; Herrero and Dickinson 1979; Herrero and Hormaza 1996). In spite of these advantages, there have been evolutionary reversals to apocarpy and/or unicarpellate condition (Endress 2011; Prychid et al. 2011; Wang et al. 2011). Nearly 10 % of the angiosperms have apocarpous gynoecia and 10 % are unicarpellate (Endress 1994, 2001). Apocarpy has greater preponderance in basal eudicots (Endress and Igersheim 2000; Endress and Doyle 2009). In higher eudicots and monocots, apocarpy is thought to be secondarily derived (Endress et al. 1983). As the mechanism of angiospermy has largely evolved independently from syncarpy (carpel fusion) (Williams et al. 2010), studies on core eudicots with unicarpellate or apocarpous gynoecia may provide fresh insights into the evolution and modularity in the structure and function of the progamic phase.

Elaeagnaceae, a core eudicot family nested in Rosales, exhibits apocarpous/unicarpellate condition and shares this feature with its sister family Barbeyaceae (Dickinson and Sweitzer 1970; Thulin et al. 1998). In its related families Rhamnaceae and Dirachmaceae, syncarpy is a prevalent condition. In Colletia spinosissima (*Rhamnaceae*), an extragynoecial compitum is present in the form of stigmatic exudate (Medan and Basilio 2001). Elaeagnaceae is represented by three genera: Hippophae, Elaeagnus and Shepherdia. The structural details of the gynoecia and information on the progamic phase in these species are not available. Among these, Hippophae rhamnoides (Himalayan sea buckthorn) is an Old World species distributed from Europe to China across the Himalaya (Mabberley 2008). It is a small to medium-size dioecious shrub and is largely exploited for its berries, known as Leh berry in India, from the wild. There are attempts to bring the species under cultivation and breeding programmes for sustainable utilization. Knowledge of the progamic phase including pollen–pistil interaction would be useful for these endeavours. In the present work on H. rhamnoides, we provide a detailed account on the structural and functional aspects of its gynoecium, and demonstrate that the species exhibits several plesiomorphic features in the gynoecium and a partly internalized pollen tube path.

METHODS

Study Species

In India, *H. rhamnoides* grows naturally between 3000 and 5000 m above sea level in the State of Jammu and Kashmir, Himachal Pradesh, Uttarakhand and Sikkim (Raina et al. 2012). The study material was collected during two flowering seasons (April/May, 2010 and 2011), from two natural populations in Leh (Jammu & Kashmir): located along the Indus River at Choglamsar (34°05.236′N, 077.36.090′E) and at Sindhu Darshan (34°05.269′N, 077.36.687′E). The climate during the flowering period is generally windy and the mean ambient temperature ranges between −12 and 14 °C.

The female flowers of sea buckthorn are borne in condensed axillary and terminal racemes. Each female flower is pedicellate and the gynoecium is enveloped in two partly fused perianth lobes. Anthesis in female flowers is marked by the emergence of a yellowish-green stigma from the perianth lobes, between 09:00 and 12:00 h

Gynoecium Architecture

Freshly opened female flowers (n = 100) were dissected from randomly collected inflorescences and the morphological details of flowers were studied with the aid of a stereo-zoom microscope (Magnüs MS 24, Olympus, India). The morphometric details were recorded with the help of a calibrated ocular micrometer (Erma, Tokyo).

For anatomical details, the pollinated and unpollinated gynoecia were fixed in Karnovsky's fixative for 4 h, rinsed with freshly prepared sodium cacodylate buffer (0.2 M) twice, dehydrated in an ethanol series, and then infiltrated and embedded in a glycol methacrylate monomer mixture (Merck). The sections (2, 4 and 5 μm) were obtained using glass knives on a rotary microtome (AO Spencer, USA), stained with toluidine blue O′ (Sigma, pH 4.4; Feder and O'Brien 1968) and viewed under a photomicroscope (Primo Star, Carl Zeiss, Germany). Additionally, some sections were stained with oil red (EX1-359 or EX2-518) to localize the lipids (Lillie and Fuller 1976), auramine-O (0.01 % in 0.05 M Tris/HCl buffer, pH 7.2, EX-460 nm; EM-550 nm) for cuticle and lipids (Heslop-Harrison and Shivanna 1977), 8-ANS (0.001 % 8-anilino-1-naphthalene sulfonic acid, EX-380 nm; EM-470 nm) for proteins (Mattsson et al. 1974) and PAS (periodic acid Schiff's reagent) for polysaccharides (McGukin and Mackenzie 1958).

The Pollen Tube Path and Progamic Phase

The pollen tube path was established by localizing the tubes from manually pollinated and open-pollinated gynoecia. For manual pollinations, the most receptive stage of stigma was ascertained through the peroxidases test (Shivanna and Rangaswamy 1992). The pollinated gynoecia (n = 450) were fixed in a lactic acid–ethanol (70 %) solution (1 : 3 v/v) at intervals of 12, 20, 24, 30, 36, 42, 48, 56, 60, 70 and 72 h after pollination (hap), for 4 h at 4 °C, rinsed with 70 % ethanol and then stored in the same solution. The temporal details of tube growth in the pollinated gynoecia were established by the decolourized aniline blue staining method (Martin 1959). Observations were made under an epifluorescence photomicroscope (Eclipse 80i Nikon, Japan). The mean tube length ± standard deviation, at each time interval, was computed and plotted against the duration at which the gynoecia were chemically fixed. The average growth rate of the pollen tube was computed by dividing the mean length of pollen tubes (n = 20 pollinated gynoecia) by 48 h (i.e. 72 h minus 24 h), because the grains took nearly 24 h to show significant germination. Pollen grains with a tube nearly equal to the diameter of pollen grains were considered germinated.

Electron Microscopy

For scanning electron microscopy, gynoecia were fixed in Karnovsky's fixative (4 h, 4 °C), dehydrated using a cold acetone series (from 10 to 100 %), dried to the critical point (E-3000, Quorum Technologies, UK), sputter-coated with gold–palladium (JFC-1600 Autofine Coater, Jeol, Japan) and observed under a scanning electron microscope (JSM-6610LV-Jeol, Japan).

The ultrastructural details of the cells lining the pollen tube path were studied by transmission electron microscopy. For this, the gynoecia were fixed in Karnovsky's fixative for 4 h at 4 °C, and washed twice with sodium phosphate buffer (0.2 M). Post-fixation was done in osmium tetroxide (1 %) prepared in the buffer, for 2 h at 4 °C, dehydrated through ascending cold acetone series and embedding was done in a CY212 araldite–resin mixture. Ultrathin sections (60–90 nm) were obtained using an ultracut rotary microtome (E-Reichert, Germany), collected on carbon-coated copper grids and stained with alcoholic uranyl acetate for 10 min followed by staining with lead acetate for 8 min (Reynolds 1963). Observations were made under a transmission electron microscope (Morgagni 268D, FEI Electron Optics).

RESULTS

Structural Organization of the Gynoecium

The gynoecium of sea buckthorn is unicarpellate and the carpel exhibits a distinct dorsiventral symmetry. It is shortly stipitate and measures 3.4 ± 0.4 mm in length (n = 30) (Fig. 1A). The ventral side is marked by an apical and prominently fanned out stigma. There is a continuous groove (ventral slit) that runs downwards, half-way from the stigma to the sub-basal region of a dorsally protruded ovary (Figs 1A and 2A)

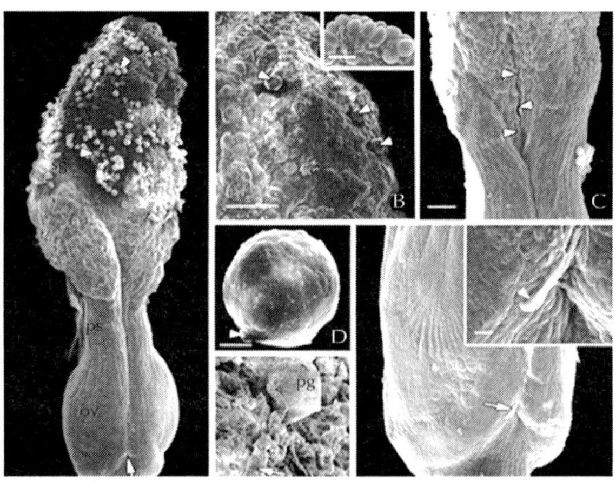

Figure 1: Scanning electron micrographs illustrating the gynoecium architecture, pollen morphology and events of progamic phase in *H. rhamnoides*. (A)

Ventral side of manually pollinated gynoecium with stipe, showing abundant pollen (arrowheads) on stigma, conduplication along the pseudostyle and ovary region which ends up in a conspicuous pore (ventral pore, arrow). (B) Part of an open-pollinated stigma showing deposition of pollen grains (arrowheads). Note that pollination causes deposition of exudates/ECM on the stigma surface due to cuticle discontinuities. Inset: Part of the unpollinated stigma at anthesis. Absence of ECM or exudates clearly indicates that the stigma is of dry type. (C) Close-up of the stigma–pseudostyle interface, and pseudostyle. Arrowheads point to the ventral slit. (D) A magnified pollen grain exhibiting smooth exine and germ pore (arrowhead). (E) Germinated pollen (30 h after pollination) showing initial phase of pollen tube growth on the stigmatic surface (arrowhead) before penetration into stigmatic tissue (arrow). (F) Ovary of a pollinated gynoecium (60 h after pollination) showing the pollen tube (arrow) near the ventral pore. The inset shows the magnified view of the pollen tube tip (arrowhead) near the pore. Abbreviations: ECM, extracellular matrix; ov, ovary; pg, pollen grain; ps, pseudostyle; sg, stigma; st, stipe. Scale bars: A = 200 μm; B = 50 μm (inset = 50 μm); C = 50 μm; D = 5 μm; E = 20 μm; F = 100 μm (inset = 10 μm).

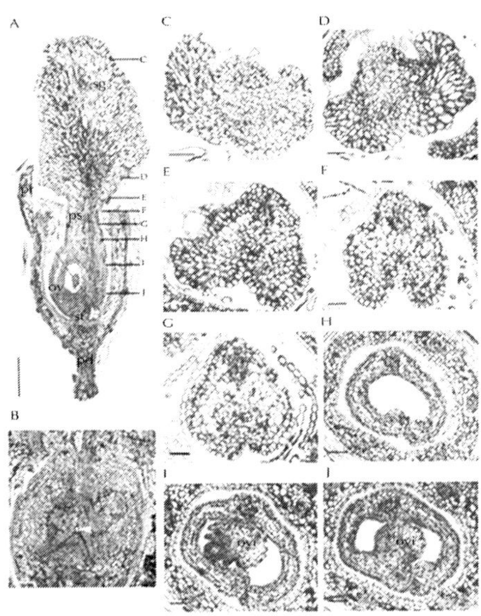

Figure 2: Anatomical details of the female flower showing the conduplicate gynoecium. (A) Longitudinal section of a young female flower showing different parts of the gynoecium. Note that the gynoecium is shortly stipitate.

The black broken line shows the plane of section of (B) of the gynoecium. (B) An oblique longitudinal section of the gynoecium from the proximal region of the gynoecium showing the ovarian pore (arrowhead) through which the pollen tube makes its entry to the ovarian locule. (C–J) Transverse sections of a young gynoecium. The corresponding regions of sections are marked in (A). The lower side of the sections represents the ventral side. (C–E) Dorsal surface of the gynoecium at the stigma region, marked by a thick cuticle (arrowhead). A solitary dorsal vascular supply is noticeable (arrow). (F and G) Pseudostyle (ps) region. A transmitting tissue (compitum) is absent. (H) Ovarian locule. Note the continuity of the locular epidermis (arrowhead) with the periphery and at the meeting point of the carpellary margins. (I) The mid-ovarian region, showing a bitegmic ovule and a developing female gametophyte. (J) The ovarian region near the pore; continuous epidermis (arrowhead) still encircles the region and shows that ovule is borne out from the submarginal position of the carpel wall. Abbreviations: ov, ovary; ovl, ovule; pe, pedicel of flower; pr, perianth; ps, pseudostyle; sg, stigma; st, stipe. Scale bars: A = 500 µm; B = 100 µm; C–E = 50 µm; F = 50 µm; G = 20 µm; H = 20 µm; I = 50 µm; J = 50 µm.

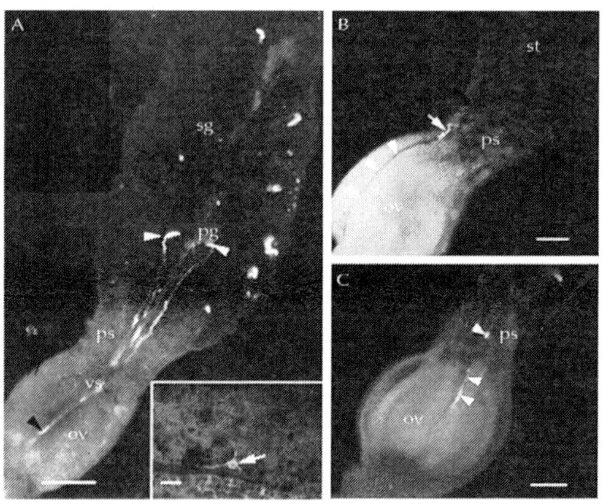

Figure 3: Epifluorescence photomicrographs of pollinated gynoecium. (A) A gynoecium 72 h after pollination showing the pollen tube along the ventral slit. Note that the tubes are directed towards the ventral slit in the pseudostyle and ovary region. The black arrowhead shows the point of entry of the pollen tube into the ovarian locule. The inset shows the pollinated gynoecium in cross-section at the ovarian region with a fluorescing pollen tube (arrow) near the ovarian pore. (B and C) A gynoecium 48 h after pollination showing a pollen tube at the stigma–pseudostyle (arrow) interface and its growth

towards the slit (arrowheads, B) and along the conduplicate zone (slit, arrowheads) of the ovarian region (C). Abbreviations: ov, ovary; pg, pollen grain; ps, pseudostyle; pt, pollen tube; sg, stigma; vg, ventral slit. Scale bars: A, B and C = 100 μm; A (inset) = 25 μm.

A freshly emerged yellowish-green stigma is the most conspicuous part of the gynoecium (2.0 ± 0.1 mm in length, n = 30), which becomes exposed to air, like a baseball glove. It is dry and of non-papillate type (Heslop-Harrison and Shivanna 1977) with an undulated (*scorbiculate*) surface (Figs 1B, 2C and D).

The stigma is subtended by a short constricted region (0.3 ± 0.1 mm in length), termed here as the pseudostyle (Figs 1A and 2A). The internalized transmitting tissue (*compitum*)-like organization is absent in this zone (Figs 2F and G). The margins of the stigma taper and merge downwards in this region to form a stigma–pseudostyle junction (Figs 1A, C and 2E). The corresponding dorsal surface of the pseudostyle is covered with dendroid trichomes (Figs 1A and 2E). The ventral slit becomes conspicuous in the pseudostyle region due to involution of the carpel margins (Fig 1A and C).

The ovary (0.8 ± 0.1 mm in length) is superior, unilocular with a single bitegmic, anatropous and crassinucellate ovule (Fig. 2A and I). The ovarian region is distinctly conduplicated as is evident by a ventral slit, which eventually leads to an ovarian pore near the base of the ovary (Figs 1A, F, 2B and 3B). The ovarian pore makes a continuum between the pollen tube path and the locule of the ovary (Figs 2H–J and 4B). The ovule develops from the submarginal position on the ventral carpellar wall of the locule (Fig. 2J). The micropyle invariably faces away from the ovarian pore (Fig. 2I).

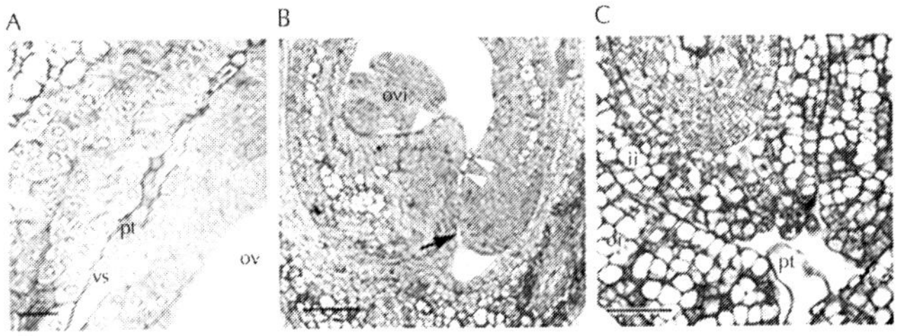

Figure 4: Pollen tube path near the ovule. (A) The pseudostyle–ovary interface with a pollen tube growing along the ventral slit. (B) A pollen tube entering

(arrowheads) through the ovarian pore. The black arrow shows the external opening of the pore. (C) Longitudinal section of the ovule showing pollen tube entry into the micropyle. Abbreviations: ii, inner integument; oi, outer integument; ov, ovary; ovl, ovule; pt, pollen tube; vs, ventral slit. Scale bars = 50 μm.

The Progamic Phase and Pollen Tube Path

The pollen grains measured 26.47 ± 3.27 μm in diameter (n = 100 pollen grains, 10 flowers) and are shed from anthers at the two-celled stage the exine is smooth (Fig. 1D). On average, the pollen grains took ~20 h to produce a small protuberance from the germ pore. After 24 h of pollination, 81.88 ± 14 % of the pollen showed tube emergence nearly equal to the mean diameter of pollen grain (27.05 ± 3.01 μm, n = 18 gynoecia). Thirty hours after pollination, the tubes attained a length of 51.0 ± 5.41 μm (n = 30 gynoecia) and penetrated into the subdermal tissue (Figs 1E and 5).

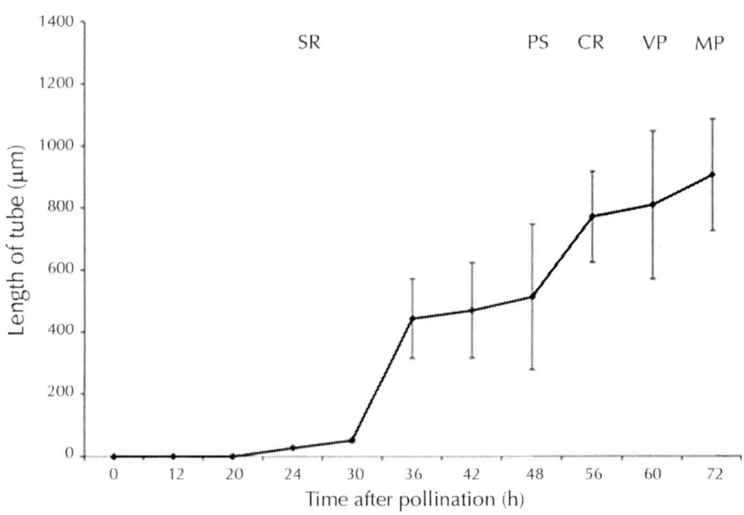

Figure 5: Temporal details of the pollen tube growth of H. rhamnoides at different intervals of time and locations in the pollinated gynoecium. The data represent the average tube length (±standard deviation). Abbreviations: CR, conduplicate region; PS, pseudostyle; SR, stigmatic region; VP, ventral pore; MP, micropyle.

The stigma attains receptivity soon after its emergence from the perianth and reaches its peak once it outgrows the subtending bract. Before pollination, the stigma is covered with a continuous cuticle, which is thicker over the median dorsal and non-receptive surface. After pollination, the cuticle lining the ventral surface becomes discontinuous (Figs 6A and 7A). Five or six layers of the subdermal cells of stigma are large, vacuolated and possess intercellular spaces. These interstitial spaces accumulate a predominantly *polysaccharidic* and *proteinaceous* extracellular matrix (ECM) (Figs 6B, 7B and C). The secretory nature of these cells was evident from the presence of numerous mitochondria and endoplasmic reticulae along the plasma membrane (Fig. 6B). The marginal cells of stigma showed accumulation of oils and *phenolics* (Fig. 7B, D and E).

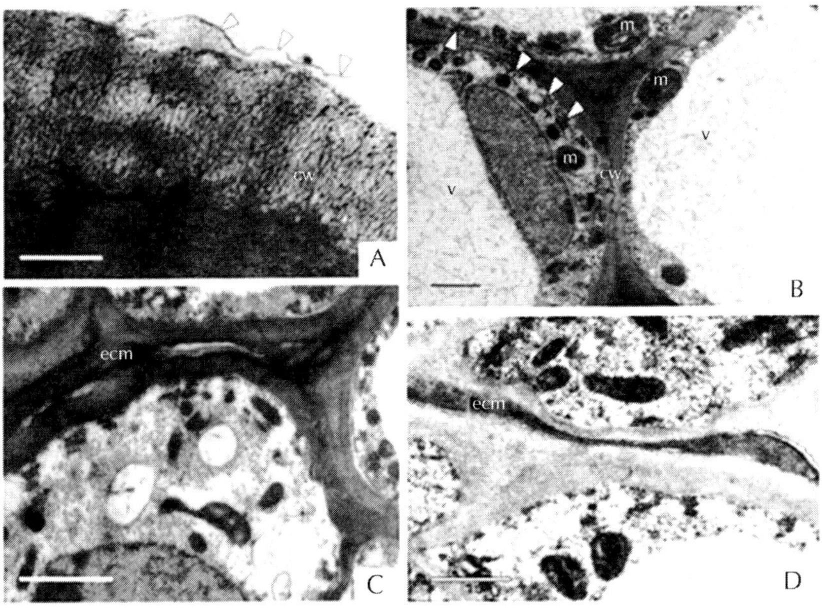

Figure 6: Transmission electron micrographs of different regions of the gynoecium. (A) A part of the stigmatic cell with an undulating cuticle (arrowheads) forming irregular bulges over the cell wall. (B) Large, vacuolated cells of stigma with intercellular spaces. The stigmatic cells contain numerous rough endoplasmic reticulae (arrowheads) and mitochondria along the plasma membrane, indicating their secretory nature. (C) Extracellular matrix accumulated

in the interstitial spaces of the hypodermal region in the pseudostyle along the site of conduplication. (D) Sparsely localized ECM between the epidermal cells lining the pore. Abbreviations: cw, thick cell wall; ecm, extracellular matrix; m, mitochondria; n, nucleus; rer, rough endoplasmic reticulae; v, vacuole. Scale bars: A = 0.5 μm; B and C = 2 μm; D = 1 μm.

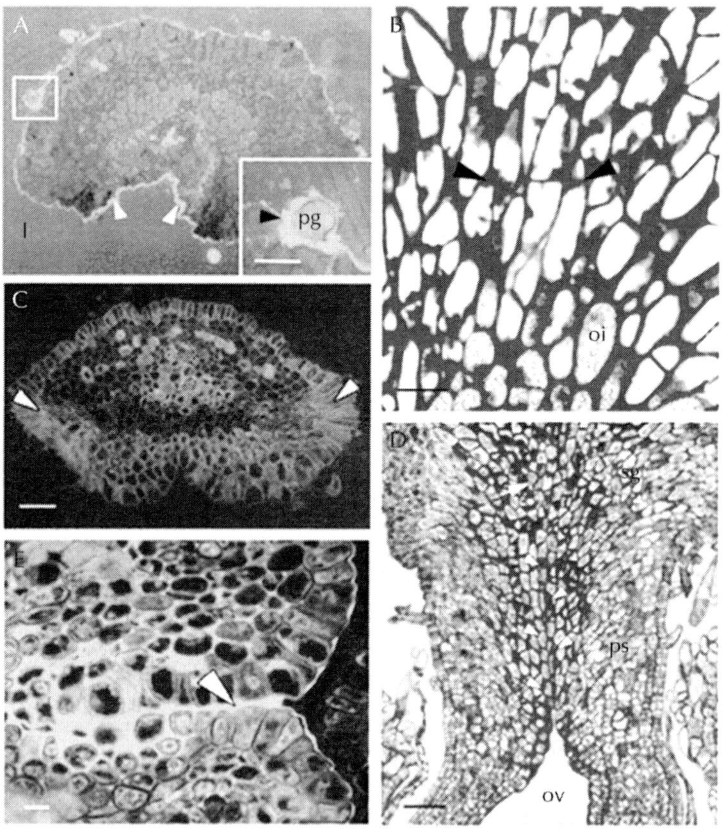

Figure 7: Histochemistry of the pollen tube path in *H. rhamnoides*. (A) Pollinated stigma, stained with auramine-O, showing the presence of cuticle. The cuticle is thicker over the dorsal surface (arrowheads). The inset shows the pollen grain submerged in the ECM deposited over the stigmatic surface, after pollination due to the development of discontinuities in the cuticle layer. (B) Stigmatic cells in the longitudinal section showing accumulation of the ECM, rich in carbohydrate (PAS positive), within the intercellular spaces (arrowheads) and the oil in some of the cells. (C) Stigma of the pollinated

gynoecium, stained with 8-ANS, showing deposition of proteins towards the ventral surface (arrowheads). The ventral surface is at the top. (D) Longitudinal section of a part of the stigma (toluidine blue O' stained), pseudostyle and upper region of the ovary showing ECM accumulation in the intercellular spaces of the stigmatic cells and along the ventral groove (arrowheads). The arrow shows one of the oil-containing cells. (E) The pseudostyle region stained with oil red, showing the presence of lipids in the ECM (arrowhead), accumulated beneath the ventral slit. Abbreviations: ECM, extracellular matrix; oi, oil. Scale bars: A = 50 μm; inset = 20 μm; B = 10 μm; C = 50 μm; D = 50 μm; E = 50 μm.

Tubes grew through the intercellular spaces of the substigmatic tissue, before reaching the stigma–pseudostyle junction, 48 hap (Figs 3A, B and 5). Interestingly, after a subdermal phase of growth, the tube(s) emerge near the junction and grow over the epidermal surface of the slit before making their entry into the locule through the ovarian pore (Figs 3, 4A and B).

Pollen tube growth along the ventral slit is supported by ECM secreted from the secretory activity of juxtaposed cells. In the pseudostyle region, the cells lining the slit are smaller, globular and compactly arranged whereas the hypodermal region possess interstitial spaces. These spaces accumulate osmiophilic and protein-rich ECM (Figs 6C, 7D and E); carbohydrates are sparsely distributed.

The pollen tube reaches the pore (Figs 1F, 3C and 4B) by 60 hap (n = 20 gynoecia) and enters the micropyle (Fig. 4C) by ~72 hap (n = 15 gynoecia). The ventral slit in the ovarian region and the ovarian pore is lined by a continuous layer of compactly arranged and thick-walled epidermis (Fig. 2H–J). A sparse osmiophilic ECM could be localized all along the ventral slit of the ovarian region and in the intervening space of the ovarian pore (Fig. 6D).

In nearly 90 % of the instances (n = 30), only one pollen tube traversed through the pseudostyle, and in the remaining 10 %, two pollen tubes were observed (Fig. 3A). In the latter case, however, invariably only one pollen tube could make its entry through the ovarian pore (Fig. 3C) and reached the micropyle. The mean pollen tube length in the pollinated gynoecia after 72 h was 908.13 ± 180.43 μm (range: 600–1300 μm, Fig. 5). Thus, the pollen tube growth rate from the time of pollen germination to ovule penetration was 18.75 μm h−1.

DISCUSSION

This is the first comprehensive account of the structural and functional aspects of the progamic phase in any taxa of *Elaeagnaceae*. Many character states in H. rhamnoides are plesiomorphic, as they have preponderance in basal angiosperms and the basal eudicots, and provide new insight into the pollen tube path.

Structural Organization of the Gynoecium

The gynoecial features in sea buckthorn, such as conduplication, sparse exudation of ECM along the site of angiospermy and incomplete fusion of carpel margins, together strongly suggest that angiospermy in sea buckthorn is contributed mainly by secretion and to some extent by the closely appressed carpel margins (Lloyd and Wells 1992; Endress and Igersheim 2000). The condition is close to type 2 angiospermy, where carpel closure is achieved by secretion and *postgenital fusion* (Endress and Igersheim 2000). The intervening constricted region between the stigma and ovary lacks the usual specialized transmitting tissue (*compitum*), and hence is termed a pseudostyle rather than a style (Bailey and Swamy 1951).

More than half of the ventral (*adaxia*l) surface is receptive (*stigma*). This is in contrast to some extant basal angiosperms with style-less, conduplicated carpels where the receptive secretory/stigmatic hairs extend to the inner surface of the carpel and pollen grains are received on two independent papillate margins known as stigmatic crests (Bailey and Swamy 1951; Endress 2005). During the evolutionary modification of the carpel, deformation of either the abaxial or adaxial surface is instrumental in displacing the stigma from a stigmatic crest to the terminal region of the style (Bailey and Swamy 1951). In H. rhamnoides, adaxial deformation seems to have developmentally organized the stigma.

An incomplete connation of carpel margins in sea buckthorn is also evident from the continuation of their epidermis into the ovarian locule, which leaves a ventral pore in the lower end of the slit. In taxa such as Kadsura, Schisandra (*Schisandraceae*) and Sagittaria (*Alismataceae*), the ventral pore may serve as the exit as well the entry point of pollen tubes, in a continuum of *extragynoecial compitum* (Lyew et al. 2007).

In *H. rhamnoides*, the ventral pore is the only route taken by the pollen tubes to gain access to the ovule.

The formation of a hunchback-like and dorsally protruded ovary in sea buckthorn is largely a consequence of the syntropous orientation of anatropous ovule, which is most likely accompanied by conduplication to accommodate the more or less basally positioned ovule. This causes the micropyle of the ovule to face away from the ventral pore. A similar condition is seen where an ovule borne on the submarginal position in free carpels is curved away from the margins in the direction of involution (Endress 1994). The dorsal bulge in the ovary around the ovule is considered an evolutionary specialization during carpel development, as in some taxa of *Monimiaceae* with a weakly peltate carpel and a *dorsal bulge* (Endress 2011).

The Progamic Phase and Pollen Tube Path

Sea buckthorn is a wind-pollinated plant and inhabits ecological conditions suitable for anemophily. The floral contrivances such as dry stigma, copious production of dry, starchy pollen with smooth exine and their morphometric range are similar to a majority of *anemophilous taxa* (Culley et al. 2002). Further, a large stigma in *H. rhamnoides* is also an obvious adaptive attribute to capture the airborne pollen grains.

During the functional phase of a dry stigma, the cuticle–pellicle layer is usually complete (Heslop-Harrison and Shivanna 1977). The cuticle layer develops discontinuities after pollination, which is accompanied by accumulation of secretion products beneath the cuticle–pellicle layer. This build-up of stigmatic ECM results in the formation of irregular bulges over the stigmatic surface, as seen in other systems with dry stigma (Shivanna and Johri 1985). The stigmatic ECM available through the cuticular discontinuities possibly facilitates adhesion and hydration of airborne pollen grains of species with dry stigma (Elleman et al. 1992).

Except for the fact that the species lacks an *extragynoecial compitum* by virtue of having a solitary and unicarpellate gynoecium, the partly hidden pollen tube growth pattern is similar to that reported in *Winteraceae* (Frame 2003), Schisandraceae (Lyew et al. 2007) and the other basal angiosperms (Endress and *Igersheim* 2000; Koehl 2002; Bernhardt et al. 2003; Thien et al. 2003; Lora et al. 2010). Enfolding of the carpel margins in the pseudostyle offers protection to the growing pollen tube (Lyew et al. 2007) and to a minor amount of ECM present over the secretory slit. The role of the pseudostyle is more or less similar to that of the usual compitum in confining the pollen tube growth along the secretory path and regulating the number of pollen tubes (Erbar 2003). In *Annona cherimola* and *Amborella trichopoda,* the stigma–style interface has been described as the site where male gametophyte selection and reduction in the number of pollen tubes occur; and only one or two pollen tubes enter the stylar canal (Williams 2009; Lora et al. 2010). In *H. rhamnoides,* male *gametophyte* competition is possibly confined to the stigmatic region and those with greater vigour nearly reach the pseudostyle. However, further down, the most vigorous one is able to grow beyond the stigma–pseudostyle interface due to spatial constraints (Erbar 2003) and thus in this context the interface appears to serve as the site of regulation of tube number.

The pollen tube growth rate is species specific and may be influenced by the length of the pollen tube path and ambient temperature regimen (Williams 2008, 2009; Steinacher and Wagner 2012). The average growth rate of the pollen tube in H. rhamnoides (18.75 μm h−1) is comparable to that of gymnosperms such as *Ephedra distachya* (14 μm h−1) and E. trifurca (6–19 μm h−1) (Williams 2009). In H. rhamnoides, the slow metabolic growth due to sub-zero temperature could be the cause of a slower tube growth rate.

CONCLUSIONS

In the evolutionary history of flowering plants, several ancestral reproductive features in the relatively advanced taxa are believed to be a consequence of reversals (Doyle and Endress 2000; Endress and Doyle 2009; Endress 2011). The presence of many *plesiomorphic* traits in the gynoecia of derived lineages suggests *homoplasic apomorphy* either by reversion or secondary derivation (Endress 1982), which is most likely influenced by the fitness requirements under the prevailing ecological conditions of a plant species (Armbruster et al. 2002). Additionally, many features such as open habitats, dioecy, simple and small flowers, uniovulate condition and wind pollination have been shown to exhibit correlated evolution (Friedman and Barrett 2008). The occurrence of *conduplication,* an incomplete connation of the carpellar margins resulting in a ventral pore and a partly internal progamic,phase represent a peculiar combination of gynoecial features in H. *rhamnoides,* a core eudicot nested within Rosales (APG III 2009). It is likely that, together, these features are the derived ones that have accompanied modification of the gynoecium of sexually diverging individuals in a dioecious system.

REFRENCES

1. APG III. An update of the Angiosperm Phylogeny Group classification for the orders and families of flowering plants: APG III. Botanical Journal of the Linnaean Society 2009;161:105-121.
2. CrossRefWeb of Science
3. Armbruster WS, Debevec EM, Willson MF. Evolution of syncarpy in angiosperms: theoretical and phylogenetic analyses of the effects of carpel fusion on offspring quantity and quality. Journal of Evolutionary Biology 2002;15:657-672.
4. Bailey IW, Swamy BGL. The conduplicate carpel of dicotyledons and its initial trends of specialization. American Journal of Botany 1951;38:373-379.
5. Bell PR. Incompatibility in flowering plants: adaptation of an ancient response. The Plant Cell 1995;7:5-16.
6. Bernhardt P, Sage T, Weston P, Azuma H, Lam M, Thien LB, Bruhl J. The pollination of Trimenia moorei (Trimeniaceae): floral volatiles, insect/wind pollen vectors and stigmatic self-incompatibility in a basal angiosperm. Annals of Botany 2003;92:445-458.
7. Culley TM, Weller SG, Sakai AK. The evolution of wind pollination in angiosperms. Trends in Ecology and Evolution 2002;17:361-369.
8. Dickinson W, Sweitzer EM. The morphology and relationships of Barbeya oleoides. American Journal of Botany 1970;57:468-476.
9. Doyle JA, Endress PK. Morphological phylogenetic analysis of basal angiosperms: comparison and combination with molecular data. International Journal of Plant Sciences 2000;161:121-153.
10. Elleman CJ, Franklin-Tong V, Dickinson HG. Pollination in species with dry stigmas: the nature of the early stigmatic response and the

pathway taken by pollen tubes. New Phytologist 1992;121:413-424.

11. Endress PK. Syncarpy and alternative modes of escaping: disadvantages of apocarpy in primitive angiosperms. Taxon 1982;31:48-52.
12. Endress PK. Diversity and evolutionary biology of tropical flowers. Cambridge, UK: Cambridge University Press; 1994.
13. Endress PK. Origins of flower morphology. Journal of Experimental Zoology Part B: Molecular and Developmental Evolution 2001;291:105-115.
14. Endress PK. Carpels of Brasenia (Cabombaceae) are completely ascidiate despite a long stigmatic crest. Annals of Botany 2005;96:209-215.
15. Endress PK. Evolutionary diversification of the flowers in angiosperms. American Journal of Botany 2011;98:370-396.
16. Endress PK, Doyle JA. Reconstructing the ancestral angiosperm flower and its initial specializations. American Journal of Botany 2009;96:22-66.
17. Endress PK, Igersheim A. Gynoecium structure and evolution in basal angiosperms. International Journal of Plant Sciences 2000;161:S211-S223.
18. Endress PK, Jenny M, Fallen ME. Convergent elaboration of apocarpous gynoecia in higher advanced dicotyledons (Sapindales, Malvales, Gentianales). Nordic Journal of Botany 1983;3:293-300.
19. Erbar C. Pollen tube transmitting tissue: place of competition of male gametophytes. International Journal of Plant Sciences 2003;164:S265-S277.
20. Feder N, O'Brien TP. Plant microtechniques: some principles and new methods. American Journal of Botany 1968;55:123-142.
21. Frame D. The pollen tube pathway in Tasmannia insipida (Winteraceae): homology of the male gametophyte conduction tissue in angiosperms. Plant Biology 2003;5:290-296.
22. Friedman J, Barrett SCH. A phylogenetic analysis of the evolution of wind pollination in the angiosperms. International Journal of Plant Sciences 2008;169:49-58.

23. Herrero M, Dickinson HG. Pollen–pistil incompatibility in Petunia hybrida: changes in the pistil following compatible and incompatible intra-specific crosses. Journal of Cell Sciences 1979;36:1-18.

24. Herrero M, Hormaza JI. Pistil strategies controlling pollen tube growth. Sexual Plant Reproduction 1996;9:343-347.

25. Heslop-Harrison J. Male gametophyte selection and the pollen–stigma interaction. In: Mulcahy DL, editor. Gamete competition in plants and animals. Amsterdam: North Holland; 1975. p. 177-190.

26. Heslop-Harrison J. Incompatibility and the pollen–stigma interaction. Annual Review of Plant Physiology 1975;26:403-425.

27. Heslop-Harrison Y, Shivanna KR. Receptive surface of angiosperm stigma. Annals of Botany 1977;41:1233-1258.

28. Koehl V. Toronto, Canada: University of Toronto; 2002. Functional reproductive biology of Illicium floridanum (Illiciaceae). MS Thesis.

29. Labarca C, Loewus F. The nutritional role of pistil exudate in pollen tube wall formation in Lilium longiflorum. II. Production and utilization of exudate from stigma and stylar canal. Plant Physiology 1973;52:87-92.

30. Lillie RD, Fuller HM. Histopathologic technique and practical histochemistry. New York: McGraw-Hill; 1976.

31. Linskens FH. Recognition during the progamic phase. In: Cresti M, Dallai R, editors. Symposium on the biology of reproduction and cell motility in plants and animals. Siena: University of Siena; 1986. p. 21-31.

32. Lloyd DG, Wells MS. Reproductive biology of a primitive angiosperm, Pseudowintera colorata (Winteraceae) and the evolution of the pollination systems in the Anthophyta. Plant Systematics and Evolution 1992;181:77-95.

33. Lora J, Hormaza JI, Herrero M. The progamic phase of an early-divergent angiosperm Annona cherimola (Annonaceae). Annals of Botany 2010;105:221-231.

34. Lyew J, Li Z, Liang-chen Y, Yi-bo L, Sage TL. Pollen tube growth in association with a dry-type stigmatic transmitting tissue and

extragynoecial compitum in the basal angiosperm Kadsura longipedunculata (Schisandraceae). American Journal of Botany 2007;94:1170-1182.

35. Mabberley DJ. A portable dictionary of plants, their classifications, and uses. 3rd edn. Cambridge, UK: Cambridge University Press; 2008.
36. Martin FM. Staining and observing pollen tubes in the style by means of fluorescence. Stain Technology 1959;34:436.
37. Mattsson O, Knox RB, Heslop-Harrison J, Heslop-Harrison Y. Protein pellicle of stigmatic papillae as a probable recognition site in incompatibility reactions. Nature 1974;247:298-300.
38. McGukin WE, Mackenzie BF. An improved periodic acid fuschin sulphite staining method for evaluation of glycoproteins. Clinical Cytochemistry 1958;4:476-483.
39. Medan D, Basilio AM. Reproductive biology of Colletia spinosissima (Rhamnaceae) in Argentina. Plant Systematics and Evolution 2001;229:79-89.
40. Prychid CJ, Sokoloff DD, Remizowa MV, Tuckett RE, Yadav SR, Rudall PJ. Unique stigmatic hairs and pollen-tube growth within the stigmatic cell wall in the early-divergent angiosperm family Hydatellaceae. Annals of Botany 2011;108:599-608.
41. Raina SN, Jain S, Sehgal D, Kumar A, Dar TH, Bhat V, Pandey V, Vaishnavi S, Bhargav A, Singh V, Rani V, Tandon R, Tewari M, Mahmoudi A. Diversity and relationships of multipurpose sea buckthorn (Hippophae L.) germplasm from the Indian Himalayas as assessed by AFLP and SAMPL markers. Genetic Resources and Crop Evolution 2012;59:1033-1053.
42. Reynolds ES. The use of lead citrate at high pH as an electron-opaque stain in electron microscopy. The Journal of Cell Biology 1963;17:208-212.
43. Sage TL, Hristova-Sarkovski K, Koehl V, Lyew J, Pontieri V, Bernhardt P, Weston P, Bagha S, Chiu G. Transmitting tissue architecture in basal-relictual angiosperms: implications for transmitting tissue origins. American Journal of Botany 2009;96:183-206.
44. Scutt CP, Vinauger-Douard M, Fourquin C, Finet C, Dumas C. An evolutionary perspective on the regulation of carpel development. Journal of Experimental Botany 2006;57:2143-2152.

45. Shivanna KR, Johri BM. The angiosperm pollen. Structure and function. New Delhi: Wiley Eastern Limited; 1985.
46. Shivanna KR, Rangaswamy NS. Pollen biology: a laboratory manual. Berlin: Springer; 1992.
47. Steinacher G, Wagner J. Effect of temperature on the progamic phase in high-mountain plants. Plant Biology 2012;14:295-305.
48. Thien LB, Sage TL, Jaffre T, Bernhardt P, Pontieri V, Weston PH, Malloch D, Azuma H, Graham SW, McPherson MA, Rai HS, Sage RF, Dupre JL. The population structure and floral biology of Amborella trichopoda (Amborellaceae). Annals of the Missouri Botanical Garden 2003;90:466-490.
49. Thulin M, Bremer B, Richardson J, Niklasson J, Chaei FF, Chase MW. Family relationships of the enigmatie rosid genera Barbeya and Dirachma from the Horn of Africa region. Plant Systematics and Evolution 1998;213:103-119.
50. Wang Xiao-Fan, Armbruster WS, Shuang-Quan H. Extra-gynoecial pollen-tube growth in apocarpous angiosperms is phylogenetically widespread and probably adaptive. New Phytologist 2011;193:253-260.
51. Williams JH. Novelties of the flowering plant pollen tube underlie diversification of a key life history stage. Proceedings of the National Academy of Sciences of the USA 2008;105:11259-11263.
52. Williams JH. Amborella trichopoda (Amborellaceae) and the evolutionary developmental origins of the angiosperm progamic phase. American Journal of Botany 2009;96:144-165.
53. Williams JH, McNeilage RT, Lettre MT, Taylor ML. Pollen tube growth and the pollen-tube pathway of Nymphaea odorata (Nymphaeaceae). Botanical Journal of the Linnean Society 2010;162:581-593.

Chapter 8

Genetic Diversity in Napier Grass (Pennisetum Purpureum) Cultivars: Implications for Breeding and Conservation

Bramwel W. Wanjala[1,2,3], Meshack Obonyo[2], Francis N. Wachira[2,7], Alice Muchugi[4], Margaret Mulaa[5], Jagger Harvey[3], Robert A. Skilton[3], Janice Proud[6], and Jean Hanson[6]

[1]Biotechnology and Biodiversity, Kenya Agricultural Research Institute, PO Box 14733, Nairobi 00800, Kenya

[2]Biochemistry and Molecular Biology Department, Egerton University, PO Box 563, Egerton 20115, Kenya

[3]Biosciences Eastern and Central Africa (BecA) at ILRI Hub, PO Box 30709, Nairobi 00100, Kenya

[4]International Centre for Research in Agroforestry, World Agroforestry Centre, United Nations Avenue, Gigiri, PO Box 30677, Nairobi 00100, Kenya

[5]Kenya Agricultural Research Institute, PO Box 450, Kitale 30200, Kenya

[6]International Livestock Research Institute (ILRI), PO Box 5689, Addis Ababa, Ethiopia

[7]Present address: Association for Strengthening Agricultural Research in Eastern and Central Africa, PO Box 765, Entebbe, Uganda

ABSTRACT

Napier grass is an important forage crop for dairy production in the tropics; as such, its existing genetic diversity needs to be assessed for conservation. The current study assessed the genetic variation of Napier grass collections from selected regions in Eastern Africa and the International Livestock Research Institute Forage Germplasm-Ethiopia. The diversity of 281 cultivars was investigated using five selective amplified fragment length polymorphism (AFLP) markers and classical population genetic parameters analysed using various software. The number of bands generated was 216 with fragments per primer set ranging from 50 to 115. Mean percentage polymorphic loci was 63.40. Genetic diversity coefficients based on Nei's genetic diversity ranged from 0.0783 to 0.2142 and Shannon's information index ranged from 0.1293 to 0.3445. The Fst value obtained was moderately significant (Fst = 0.1688). Neighbour-joining analysis gave two distinct clusters which did not reflect geographical locations. Analysis of molecular variance showed all variance components to be highly significant ($P < 0.001$), indicating more variation within (91 %) than between populations (9 %). Results suggested moderate genetic differentiation among Napier grass populations sampled, which could imply a high germplasm exchange within the region. The AFLP markers used in this study efficiently discriminate among cultivars and could be useful in identification and germplasm conservation.

INTRODUCTION

In East Africa, Napier grass (*Pennisetum purpureum*) is a perennial grass grown widely as a fodder crop and feed for the cut-and-carry zero-grazing dairy systems (Bayer 1990) and constitutes up to 80 % of forage for smallholder dairy farms (Staal et al. 1987). It is the forage of choice not only in the tropics but also worldwide (Hanna et al. 2004) due to its desirable traits such as tolerance to drought and a wide range of soil conditions, and high photosynthetic and water-use efficiency (Anderson et al. 2008).

While much attention has been directed towards research for improving the productivity of major cereal crops (Katz 2003), there has been comparatively little effort to improve Napier grass—an important forage crop that has been grown over centuries and currently enjoys a multiplicity of uses besides conventional animal consumption (Jaradat 2010). This is key among the drivers of renewed research interest in this otherwise previously neglected crop. However, the productivity of Napier grass is limited by several factors especially emerging diseases, mainly Napier grass stunt disease and Napier grass head smut disease, which constrains the growth of the smallholder dairy industry (New Agriculturists 2009). For this reason, it is necessary to strengthen forage breeding programmes for the development of disease-resistant cultivars (Boa et al. 2005).

Correct identification of Napier grass accessions is a prerequisite because the existing *germplasm* information is scanty and cannot be relied upon for crop improvement, since cultivar discrimination has predominantly relied on morphological and agronomic features and is the major cause of inconsistency in identification. Consequently, a number of Napier grass cultivars have been in circulation, often with more than one name (Struwig 2007).

Molecular markers have proven useful in distinguishing among morphologically related individuals within cultivars of the same plant species (Mohammadi and Prasanna 2003). Thus the genetic assessment of various Napier grass accessions from the Eastern Africa region is important for correct cultivar identification in order to exploit them fully in crop improvement strategies.

However, in genetic diversity studies, a wide range of marker systems require development in order to be used either singly or in combination, depending on the researcher's interest. Thus, most markers have some degree of shortcomings, which include: (i) restriction fragment length polymorphisms (RFLPs) are time consuming and involve expensive radioactive materials (Mondini et al. 2009), (ii) random amplified polymorphic DNAs (RAPDs) are irreproducible (Bardakci 2001), (iii) isozyme systems are few per species, (iv) amplified fragment length polymorphism (AFLP) band profiles cannot be interpreted in terms of loci, alleles and their dominance (Spooner et al. 2005), and (v) there is a lack of simple sequence repeats (SSRs) for Napier grass (Azevedo et al. 2012). Thus, due to a large number of loci analysed, high polymorphism levels, amenability to automation, high reproducibility without prior sequence knowledge and genome-wide marker distribution (Powell et al. 1996), the AFLP method (Vos et al. 1995) is considered more reliable and robust for an evaluation of genetic variability.

This study assessed the genetic variation between and within Napier grass collections comprising 281 accessions from selected regions in Eastern Africa (Kenya, Uganda, Tanzania and International Livestock Research Institute Forage Germplasm-Ethiopia). Implications of the findings for germplasm identification, breeding and conservation are discussed.

METHODS

Source of Cultivars

Cultivars were collected from selected regions in Kenya, Uganda and Tanzania between 2009 and 2010. Also included was a collection from the International Livestock Research Institute Forage Germplasm (ILRI-FG) which comprised cultivars from other parts of Africa, USA and accessions of unknown origin. The ILRI-FG was established in 1982 and has been maintained without external infusion (Table 1). Samples were selected based on morphological differences and collected from different regions with a range of ecological and altitudinal variations. They were then evaluated for nutritive parameters and biomass forage yield compared with the best local clone at 8 weeks post-harvest intervals. From the 281 clones, 3 young leaves without necrosis were collected and placed in a transparent polythene bag. Holes were punched to permit air flow. Samples were dried by placing them in a large plastic bag with silica. The silica gel was changed daily until samples were completely dry. The dried samples were then packaged in large polythene bags (without holes to prevent rehydration) and sent to the Biosciences east and central Africa (BecA)-ILRI Hub, Nairobi for analysis.

Table 1: Genetic diversity indices. Population, sub-population, sample size (N), mean diversity estimates (H) and Shannon's information index (I), polymorphic loci and % polymorphic loci. *Samples from Ethiopia = 2, Nigeria = 2 and Ghana = 2.

Populations	Sub-populations	N	H	I	No. of polymorphic loci	°/0 Polymorphic loci
Kenya	Bungoma	27	0.1169	0.2076	167	77.3
	Busia	25	0.1197	0.2095	154	71.3
	Butere	22	0.1492	0.2569	187	86.6
	Alupe	11	0.1195	0.2071	140	64.8
	Mumias	35	0.1416	0.2464	193	89.4
Tanzania	Hai	5	0.1747	0.2675	116	53.7
	Lushoto	4	0.1727	0.2633	111	51.3
	Meru	3	0.1608	0.2390	92	42.6
	Iluheza	14	0.1011	0.1771	128	59.3
	Tarime	19	0.1893	0.3113	186	83.1
Uganda	Kabarole	12	0.2130	0.3353	168	77.8
	Masaka	23	0.1244	0.2099	151	69.9
	Naro	5	0.0167	0.0266	13	32.5
	Soroti	16	0.2142	0.3445	189	87.5
ILRI-FG	Others	6*	0.0783	0.1293	71	32.9

	Swaziland	6	0.1576	0.2495	123	56.9
	Tanzania	5	0.1308	0.2035	93	43.1
	Unknown	14	0.1338	0.2258	145	67.1
	USA	16	0.1890	0.3044	170	78.7
	Zimbabwe-C	8	0.1844	0.2938	152	70.4

DNA Extraction

Genomic DNA was extracted following the cetyltrimethylammonium bromide (CTAB) method (Doyle and Doyle 1987) with some modifications. Dried leaves of ~200–300 mg were cut into small pieces and transferred into 1.2-mL strip tubes containing two stainless steel grinding balls. Strip tubes were then cooled by immersing in liquid nitrogen. The frozen leaves were subsequently ground into fine powder using GenoGrinder-2000 at a speed of 500 strokes min−1 for 10 min. Strip tubes were centrifuged at 1000 rpm for 1 min to allow collection of the ground tissue at the bottom. Warm CTAB buffer (700 μL; 65 °C) (2 % CTAB, 100 mM Tris–HCl pH 8.0, 1.4 M NaCl, 20 mM EDTA, 2 % polyvinylpyrrolidone (PVP)) was added and the samples homogenized for 1 min by re-grinding in a GenoGrinder. Samples were incubated in a water bath at 65 °C for an hour with continuous shaking and mixing by inversion of strip tubes (every 15 min). Tubes were then removed from the water bath and left to cool at room temperature for 10 min, followed by centrifugation at 4000 rpm for 20 min. The aqueous layer was transferred into fresh strip tubes. Chloroform : isoamyl alcohol (24 : 1; 600 μL) was added and the contents mixed by gently inverting the tubes 10 times. Centrifugation was carried out at 4000 rpm for 20 min and the aqueous (top) layer transferred into a fresh strip tube. The above step was repeated twice and the aqueous layer transferred into fresh strip tubes. Ice-cold absolute ethanol: sodium acetate (25 : 1; 600 μL) was added gently and mixed by inversion, and a sample incubated at −20 °C for 45 min. DNA was pelleted by centrifugation at 4000 rpm for 20 min. The supernatant was removed; the pellet was washed with 500 μL of 70 % ethanol and left to stand for 5 min, then washed twice. The DNA pellet was air dried for 20 min, then dissolved in 100 μL of low-salt TE buffer (10 mM Tris–HCl pH 7.5, 1 mM EDTA). To free the DNA of RNA, 5 μL of ribonuclease A (RNase A), 10 mg/mL, were added and incubated at 37 °C for 30 min. DNA was re-precipitated by adding 200 μL of cold absolute ethanol and incubated at −20 °C for 30 min. DNA was pelleted by centrifuging at 4000 rpm for 20 min and washed with 70 % ethanol twice. The pellet was air dried for 30–60 min and resuspended in 100 μL of low-salt TE buffer.

AFLP Analysis

The AFLP protocol described by Vos et al. (1995) was employed. Sixty-four primer pairs were initially screened for their potential to produce scorable fragments (Applied Biosystems, Foster City, CA, USA). From 64 primer pairs, five were chosen based on their reproducibility and levels of fragment polymorphism (Table 2). Reproducibility was assessed by duplicating the same sample three times with different primer pair combinations to produce similar electropherograms. Polymerase chain reaction (PCR) products were prepared for separation on capillary electrophoresis (Applied Biosystems). Size separation of AFLP–PCR fragments was done on an ABI 3730 xl Genetic Analyzer (Applied Biosystems). Amplified fragment length polymorphism peaks from the Genetic Analyzer were sized and alleles scored with GeneMapper version 4.1 (Applied Biosystems AFLP Plant Mapping Protocol, 2005). Scored results were exported to an Excel matrix with values 1 (allele present) or 0 (allele absent)

Table 2: Selected selective AFLP primer combinations

Primer no.	EcoRI primer	MseI primer	Dye	Color
1	EcoRI + AAC	MseI + CAT	NED	Yellow
2	EcoRI + ACC	MseI + CAA	NED	Yellow
3	EcoRI + ACT	MseI + CAG	6-FAM	Blue
4	EcoRI + ACT	MseI + CAT	6-FAM	Blue
5	EcoRI + AGG	MseI + CAC	JOE	Green

Data Analyses

Genetic diversity, ordination analysis and analysis of molecular variance (AMOVA)—for diversity among and within populations—were determined using GenAlEx software (Peakall and Smouse 2009). PopGen32 (Yeh et al. 1997) was used for the population-based approach using F statistics, gene diversity over loci, proportion of polymorphic loci, Shannon index and gene frequency (Nei 1987; McDermott and McDonald 1993). In addition, the genetic distance between any two populations (Schneider et al. 2000) was computed using Tools for Population Genetic Analysis (TFPGA) software (Miller 1997). To show the relationships between 281 cultivars, principal co-ordinate analysis and unweighted pair group method with arithmetic mean (UPGMA) were generated using Darwin software (Perrier and Jacquemoud-Collet 2006). About 2000 bootstrap replicates were used to determine branch support in the consensus tree.

RESULT

Evaluation of Genetic Diversity

From the initial 64 primers screened, 22 were efficient at the inter-population level and sufficiently polymorphic to discriminate clones within populations. Intra-population polymorphism obtained with the selected primers gave alleles unique to each individual from the same population. Five primers that generated electropherograms with high relative fluorescent units/(peaks) without background noise were further selected for genotyping. Amplified fragment length polymorphism fragments ranged from 50 to 500 base pairs. Polymorphic bands ranged from 50 to 115 with an average of 43 bands per primer from a total of 216 bands generated, accounting for 64.80 % of polymorphic loci observed. The percentage of polymorphic fragments within sub-populations ranged from 32.5 for the National Agricultural Research Organization (NARO) to 89.4 % for Mumias. Following the same order, the genetic diversity coefficients based on Nei's genetic diversity ranged from 0.0783 to 0.2142 and Shannon's information index ranged from 0.1293 to 0.3445 (Table 1).

Phenetic Analysis

The genetic distances between/among 21 Napier grass sub-populations (regions within countries) and four populations (countries: Kenya, Uganda, Tanzania and ILRI-FG) were subjected to hierarchical clustering by UPGMA (Table 3). This yielded two distinct clusters (A and B) which did not reflect the geographical locations of the 281 cultivars (Fig. 1). In addition, there was an overlap among cultivars spread across different clusters (Fig. 2). Pairwise comparison of genetic distance and similarity of populations revealed little genetic diversity within the Kenyan population and the ILRI-FG population, while it was moderate within the Uganda and Tanzanian populations

Table 3. Nei's genetic distance of the 21 sub-populations of Napier grass based on AFLP analysis. Population codes: 1 = Bungoma, 2 = Busia, 3 = Butere, 4 = Extra, 5 = Hai, 6 = Kabarole, 7 = Lushoto, 8 = Masaka, 9 = Meru, 10 = Muhenza, 11 = Mumias, 12 = Naro, 13 = Others, 14 = Soroti, 15 = Swaziland, 16 = Tanzania, 17 = Tarime, 18 = Unknown, 19 = USA, 20 = Zimbabwe-cultivar and Zimbabwe-hybrids. Bold values are significant.

Pop ID	1	2	3	4	5	6	7	8	9	10	11
1	****										
2	0.0006	****									
3	0.0055	0.0034	****								
4	0.0015	0.0003	0.0014	****							
5	0.0531	0.0520	0.0487	0.0667	****						
6	0.0376	0.0372	0.0336	0.0458	0.0207	****					
7	0.0897	0.0855	0.0841	0.1024	0.0141	0.0384	****				
8	0.0032	0.0046	0.0142	0.0110	0.0419	0.0313	0.0794	****			
9	0.0151	0.0128	0.0132	0.0156	0.0334	0.0364	0.0614	0.0155	****		
10	0.0050	0.0033	0.0063	0.0002	0.0741	0.0541	0.1053	0.0151	0.0180	****	
11	0.0001	0.0006	0.0027	0.0021	0.0410	0.0302	0.0756	0.0030	0.0098	0.0057	****
12	0.0108	0.0104	0.0239	0.0105	0.0971	0.0764	0.1338	0.0197	0.0261	0.0091	0.0156
13	0.0018	0.0019	0.0136	0.0073	0.0571	0.0424	0.0884	0.0028	0.0134	0.0084	0.0040
14	0.0248	0.0240	0.0236	0.0329	0.0166	0.0080	0.0262	0.0210	0.0180	0.0360	0.0189
15	0.0214	0.0205	0.0221	0.0307	0.0129	0.0114	0.0419	0.0132	0.0101	0.0379	0.0147
16	0.0252	0.0243	0.0323	0.0358	0.0223	0.0140	0.0423	0.0159	0.0245	0.0399	0.0207

17	0.0151	0.0142	0.0162	0.0220	0.0126	0.0119	0.0382	0.0103	0.0072	0.0255	0.0095
18	0.0041	0.0014	0.0061	0.0046	0.0425	0.0313	0.0639	0.0072	0.0080	0.0059	0.0023
19	0.0276	0.0262	0.0250	0.0349	0.0088	0.0159	0.0230	0.0228	0.0112	0.0391	0.0200
20	0.0237	0.0219	0.0153	0.0301	0.0130	0.0163	0.0343	0.0216	0.0026	0.0345	0.0148
21	0.0492	0.0487	0.0523	0.0645	0.0201	0.0066	0.0392	0.0349	0.0430	0.0707	0.0420

12	13	14	15	16	17	18	19	20	21
****									
0.0046	****								
0.0533	0.0261	****							
0.0502	0.0207	0.0058	****						

0.0482	0.0180	0.0091	0.0045	****					
0.0399	0.0138	0.0049	0.0020	0.0049	****				
0.0130	0.0021	0.0165	0.0152	0.0151	0.0087	****			
0.0536	0.0257	0.0036	0.0036	0.0080	0.0027	0.0148	****		
0.0554	0.0280	0.0078	0.0020	0.0194	0.0033	0.0160	0.0017	****	
0.0420	0.0431	0.0080	0.0094	0.0100	0.0165	0.0387	0.0157	0.0231	****

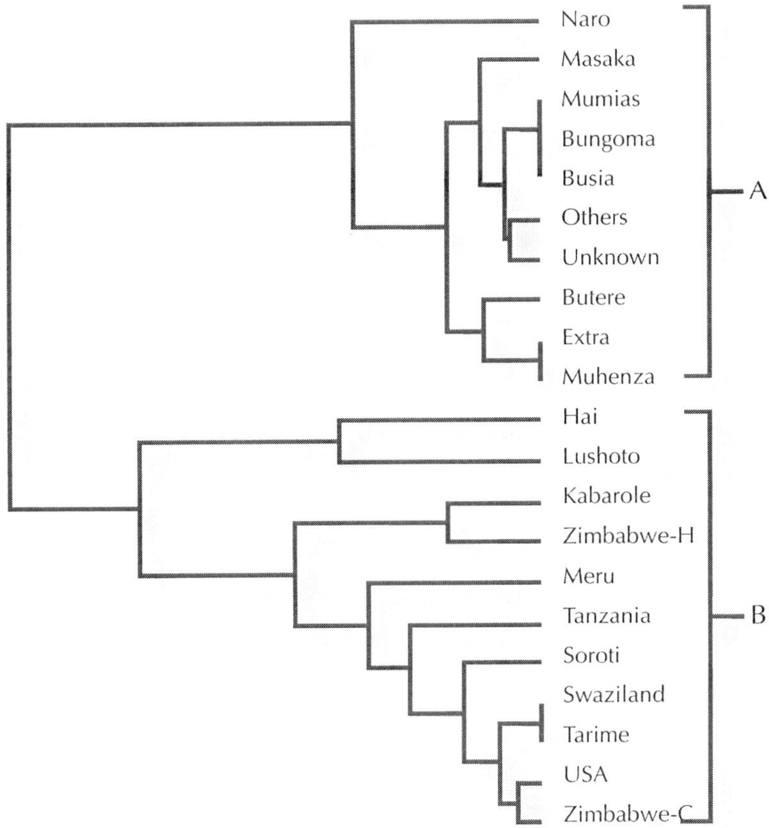

Figure 1: UPGMA dendrogram of 21 P. purpureum sub-populations from Kenya, Uganda, Tanzania and ILRI-FG using genetic distance (Nei 1979).

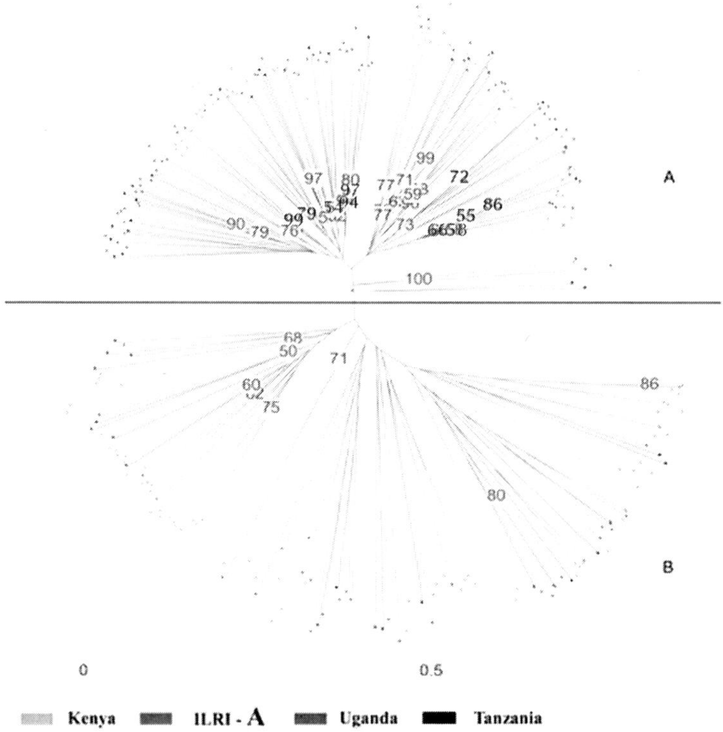

Figure 2: UPGMA neighbour-joining dendrogram of 281 Napier grass accessions computed from 216 polymorphic AFLP markers using Darwin hierarchical clustering with a bootstrap value at 2000.

Population Structure Among and Within Populations

Variance components obtained by AMOVA were highly significant (P < 0.001) with more variation within (91 %) than between populations (9 %) (Table 4). Similarly, nested analysis partitioned by country, within population, Uganda 83 %, Kenya 97 %, Tanzania 86 % and ILRI-FG 96 %, was also highly significant (P < 0.001). On the other hand, variation among population showed the Kenyan population as having the least variation (3 %) while the most variation was among the Ugandan population (17 %).

Table 4: Analysis of molecular variance for 281 genotypes from 21 Napier grass populations based on 216 AFLP markers. Df, degrees of freedom; Ss, sum of squares; Ms, mean square; Est.var, estimated variation

Source of variation All populations	Df	Ss	Ms	Est.var	Percentage	P value
Among population	20	1125.94	56.29	2.46	9	0.001
Within population	260	6249.76	24.04	24.04	91	0.001
Kenya population						
Among population	4	105.08	26.27	0.15	3	0.001
Within population	115	2611.36	22.71	22.71	97	0.001
Tanzania population						
Among population	4	231.73	57.93	4.06	14	0.001
Within population	40	1038.36	25.959	25.96	86	0.001
Uganda population						
Among population	3	273.893	91.298	5.15	17	0.001
Within population	52	1269.55	24.414	24.41	83	0.001
ILRI-FG						
Among population	6	3.018	0.503	0.001	4	0.001
Within population	53	26.10	0.492	0.49	96	0.001

DISCUSSION

Reliability of AFLP Markers

A good molecular marker must separate individuals' inter-populations and still be sufficiently polymorphic at the intra-population level to precisely identify clones (Mueller and Wolfenbarger 1999). Amplified fragment length polymorphism employed in the current study generated numerous highly polymorphic alleles, which corroborates the findings of Vos et al. (1995) that a large number of alleles are important for accurately estimating the genetic diversity of a germplasm. While microsatellites are among the most commonly used markers due to their locus specificity, co-dominant nature, high polymorphism and reproducibility, their development and application have been restricted to a few agriculturally important crops since they first require identification via genome sequencing (Powell et al. 1996). The genome of Napier grass has not been sequenced and the inputs involved are high and limit the development of microsatellite for this crop. In addition, Napier grass is a tetraploid ($2n = 4x = 28$) and triploid and hexaploid hybrids occur between it and pearl millet (Techio et al. 2010). This presents a challenge in establishing microsatellites that can clearly discriminate the different ploidy levels. The multi-locus nature of AFLP thus makes it reliable over the microsatellite procedure as it scans the entire genome.

The resolving power of a marker is an important index in selecting the most informative markers for diversity studies (Azevedo et al. 2012). For example, Harris et al. (2009) recorded a resolving power of 13.2 for AFLP while Azevedo et al. (2012) found a resolving power of 1.55 for microsatellites, implying that AFLP is more informative. The current study employed five primer pair combinations which were found to be polymorphic and sufficient in discriminating Napier grass cultivars. This is in agreement with the findings of Ellis et al. (1997) that by choosing the six best combinations of primers it would be possible to explain over 80 % of expected species relatedness; in addition Saunders et al. (2001) demonstrated that differences in AFLP-DNA fragments would be detected when no more than 3–7 primer pairs are used. Thus, this study established a reliable AFLP procedure for Napier grass.

Evaluation of Genetic Diversity

The current study was founded on the premise that Napier grass accessions could segregate based on geographical origin and hence would offer a unique genetic resource for breeding programs in the region. However, the converse appears true as there is moderate diversity among accessions in the region. The clustering suggests genotype overlap and redundancy possibly due to low gene flow. The materials under circulation share ancestry especially since Napier grass is clonally propagated, the users may be sharing planting materials. These findings contradict previous studies that demonstrated clustering of Napier grass accessions based on geographical location (Lowe et al. 2003; Harris et al. 2009). Despite the fact that Harris et al. (2009) used AFLP markers, different primer pair combinations and samples were analysed. On the other hand, Lowe et al. used different samples from those used in this study. In addition, the marker employed was RAPDS, which has several limitations, among them lack of reproducibility The low genetic distance observed among the Kenyan sub-populations may be attributed to the close proximity of the sampled areas (Bungoma, Extra-Alupe, Busia and Mumias), which are in a radius of ~60 km. Similarly, this applies to ILRI-FG, Tanzania and Uganda populations, which had moderate genetic distances. The 'Tanzania' sub-population, from ILRI-FG, clustered with Tanzanian populations, Lushoto, Tarime and Meru, indicating ancestry from Tanzania. The sub-ILRI-FG germplasm is diverse with most originating from Zimbabwe and the USA; however, their origin could not be ascertained. The USA is physically isolated from Africa, but these individuals were still flanked by the cultivars collected from Africa. This could suggest that they had not yet diversified from their African relatives. Struwig (2007) using AFLP markers on ILRI-FG genotypes observed that cultivars did not cluster based on geographical origin, which is consistent with observations of the current study.

Population Genetic Structure

Genetic structure is affected by several factors including breeding systems, genetic drift, population age and size, environmental heterogeneity, seed dispersal, gene flow, evolutionary history as well as natural selection (Hamrick and Godt 1990). This is probably attributable to the out-crossing nature of Napier grass, which has higher levels of genetic diversity and lower differentiation among populations than in selfing and clonal plants (Rossetto et al. 1995). Similar results were observed in other out-crossing species such as perennial rye grass (*Lolium perenne*), meadow fescue (*Festuca pratensis*), orchard grass (*Dactylis glomerata*) and Rhodes grass (*Chloris gayana*) (Huff 1997; Kölliker et al. 1998; Ubi et al. 2003).

The Fst values obtained between the 21 sub-populations indicate moderate genetic differentiation in Napier grass cultivars (Hartl and Clark 1989). Since the grass is a clonal plant with low seed setting and germination, it is spread by asexual stem reproduction. Therefore, gene flow among cultivars is low and most genetic variation resides between rather than within cultivars (Xie et al. 2009). Another probable reason for high within-population variation may be caused by Napier grass being a highly heterozygous tetraploid species. Variations in Napier grass cultivars are expected to be high due to its rich gene pool and wide parental diversity (Azevedo et al. 2012). Taxonomic representation is based on the cross-ability of wild species and the domesticated form and the amount of gene flow occurring (Robert et al. 1991). The genus Pennisetum has three gene pools. The primary gene pool occurs between the domestication of *Pennisetum glaucum* and wild weedy forms of *P. glaucum*. The secondary gene pool is between perennial and wild relative (*P. purpureum* to *P. glaucum*). They cross easily but their hybrids are sterile. The tertiary gene pool comprises true biological species compared with the primary and secondary gene pools (Robert et al. 1991). However, strong reproductive barriers impede natural gene flow and the occurrence of hybrids between tertiary versus other forms (Robert et al. 1991). Therefore, higher levels of genetic differentiation among cultivars could arise from the pressure of artificial selection, which occurs rarely.

Implications for Conservation and Improvement

Africa is believed to be the centre for domestication of Napier grass (Azevedo et al. 2012) as it houses a majority of the Pennisetum gene pools (Techio et al. 2010). Thus, maintaining genetic diversity within natural populations can maximize their potential to withstand and adapt to biotic and abiotic pressures (Jump et al. 2009). The possibility of interspecific combination is important for breeding as it allows the transfer of alleles into species of agronomic importance, as has been done successfully between millet and other species of Pennisetum (Techio et al. 2010).

Assessment of the entire Pennisetum gene pool would make material available for breeding programmes without causing genetic erosion or loss of varieties. Thus there ought to be renewed efforts among researchers to populate the forage germplasm (ILRI-FG) with materials from different agroecological zones of the world. Potential sources of Napier grass diversity are South Africa, Brazil, Puerto Rico, the USA, Australia, China, Pakistan and India (Azevedo et al. 2012).

REFERENCES

1. Anderson WF, Dien BS, Brandon SK, Peterson JD. Assessment of Bermuda grass and bunch grasses as feed stocks for conversion to ethanol. Applied Biochemistry and Biotechnology 2008;145:13-21.
2. Azevedo ALS, Costa PP, Machado JC, Machando MA, Pereira AV, Ledo FJS. Cross species amplification of Pennisetum glaucum microsatellite markers in Pennisetum purpureum and genetic diversity of Napier grass accessions. Crop Science 2012;4:1776-1785.
3. Bardakci F. Random amplified polymorphic DNA (RAPD) markers. Turkish Journal of Biology 2001;25:185-196.
4. Bayer W. Napier grass—a promising fodder for smallholder livestock production in the tropics. Plant Research Development 1990;31:103-111.
5. Boa E, Ajanga S, Mulaa M, Jones P. Going public on Napier stunting disease in Western Kenya. Egham, UK: Global Plant Clinic; 2005. 14 pp.
6. Doyle JJ, Doyle JL. A rapid DNA isolation procedure for small quantities of fresh leaf material. Phytochemical Bulletin 1987;19:11-15.
7. Ellis RP, McNicol JW, Baird E, Booth A, Lawrence P, Thomas B, Powell W. The use of AFLP to examine genetic relatedness in barley. Molecular Breeding 1997;3:359-369.

8. Hamrick JL, Godt MJ. Plant population genetics, breeding, and genetic resources. Sunderland, MA: Sinauer; 1990. p. 43-63.
9. Hanna WW, Chaparro CJ, Mathews BW, Burns JC, Sollenberger LE, Carpenter JR. Perennial pennisetums. In: Moser LE, Burson BL, Sollenberger LE, editors. Warm-season (C4) grasses. Madison, American Society of Agronomy: Monograph series; 2004. p. 503-535. Vol. 34.
10. Harris K, Anderson W, Malik R. Genetic relationships among Napier grass (Pennisetum purpureum) nursery accessions using AFLP markers. Plant Genetic Resources: Characterization and Utilization 2009;1:1-8.
11. Hartl DL, Clark AG. Principles of population genetics. 3rd edn. Sunderland, MA: Sinauer Associates, Inc; 1989.
12. Huff DR. RAPD characterization of heterogeneous perennial ryegrass cultivars. Crop Science 1997;37:557-594.
13. Jaradat AA. Genetic resources of energy crops: biological systems to combat climate change. Australian Journal of Crop Science 2010;4:309-323.
14. Jump AS, Marchant R, Penuelas J. Environmental change and the option value of genetic diversity. Trends in Plant Science 2009;14:51-58.
15. Katz SH. 2003. Cereal grains and pseudo-cereals. Encyclopedia of food & culture http://www.enotes.com/food-encyclopedia/cereal-grains-pseudo-cereals 14 July 2009.
16. Kölliker R, Stadelmann FJ, Reidy B, Nösberger J. Genetic variability of forage grass cultivars: a comparison of Festuca pratensis Huds. Lolium perenne L. and Dactylis glomerata L. Euphytica 1998;106:261-270.
17. Lowe AJ, Thorpe W, Teake A, Hanson J. Characterization of germplasm accession of Napier grass (Pennisetum purpureum and P. purpureum and P. glaucum hybrids) and comparison with farm clones using RAPD. Genetic Resources and Crop Evolution 2003;50:121-132.
18. McDermott JM, McDonald BA. Gene flow in plant pathosystems. Annual Review of Phytopathology 1993;31:353-373.

19. Miller MP. 1997. Tools for Population Genetic Analyses (TFPGA). Available with the program's installation files at http://www.marksgeneticsoftware.net/tfpga.htm 19 June 2009.

20. Mohammadi SA, Prasanna BM. Analysis of genetic diversity in crop plants—salient statistical tools and considerations. Crop Science 2003;43:1235-1248.

21. Mondini L, Noorani A, Pagnotta MA. Assessing plant genetic diversity by molecular tools. Diversity 2009;1:19-35.

22. Mueller UG, Wolfenbarger LL. Amplified fragment length polymorphism (AFLP) genotyping and fingerprinting. Trends in Ecology and Evolution 1999;14:389-394.

23. Nei M. Genetic distance between populations. American Naturalist 1987;106:283-292.

24. Nei M, Li WH. Mathematical model for studying genetic variation in terms of restriction endonucleases. Proceedings of the National Academy of Sciences of the USA 1979;76:5269-5273.

25. New Agriculturists. A healthy aversion to smut. http://www.new-ag.info/98-2/newsbr.html 1 November 2009.

26. Peakall R, Smouse PE. GENALEX 6: Genetic Analysis in Excel. Population genetic software for teaching and research. Molecular Ecology Notes 2009;6:288-295.

27. Perrier X, Jacquemoud-Collet JP. 2006. DARwin software http://darwin.cirad.fr/ 16 July 2009.

28. Powell W, Machray GC, Provan J. Polymorphism revealed by simple sequence repeats. Trends in Plant Science 1996;7:215-222.

29. Robert T, Lespinasse R, Peerns J, Sarr A. Gametophytic competition as influencing gene flow between wild and cultivated forms of pearl millet (Pennisetum typhoides). Genome 1991;34:195-200.

30. Rossetto M, Weaver PK, Dixon KW. Use of RAPD analysis in devising conservation strategies for the rare and endangered Grevillea scapigera. Molecular Ecology 1995;4:321-329.

31. Saunders JA, Pedroni MJ, Penrose LDJ, Fist AJ. AFLP DNA analysis of opium poppy. Crop Science 2001;41:1596-1601.

32. Schneider S, Roessli D, Excoffier L. Arlequin, version 2000: software for population genetic analysis. Switzerland: Genetics and Biometry Laboratory, University of Geneva; 2000.

33. Spooner D, van Treuren R, de Vicente MC. Molecular markers for gene bank management. Rome, Italy: International Plant Genetic Resources Institute; 2005. IPGRI Technical Bulletin No. 10.

34. Staal S, Chege L, Kenyanjui M, Kimari A, Lukuyu B, Njubi D, Owango M, Tanner J, Thorp W, Wambugu M. A cross-sectional survey of Kiambu District for the identification of target groups of smallholder dairy producers. Nairobi, Kenya; 1987. KARI/ILRI collaborative project research report.

35. Struwig M. 2007. Fingerprinting Pennisetum purpureum varieties and cultivars using AFLP analyses. Thesis, Masters of Environmental Sciences.

36. Techio VH, Davide LC, Cagliari A, Barbosa S, Pereira AV. Karyotipic asymmetry of both wild and cultivated species of Pennisetum. Bragantia Campinas 2010;69:273-279.

37. Ubi GN, Kenworthy WJ, Costa JM, Cregan PB, Alvernaz J. Genetic diversity of soybean cultivars from China, Japan, North America, and North American ancestral lines determined by amplified fragment length polymorphism. Crop Science 2003;43:1858-1867.

38. Vos P, Hogers R, Bleeker M, Reijans M, van de Lee T, Hornes M, Frijters A, Pot J, Peleman J, Kuiper M. AFLP: a new technique for DNA fingerprinting. Nucleic Acids Research 1995;23:4407-4414.

39. Xie XM, Zhou F, Zhang XQ, Zhang JM. Genetic variability and relationships between MT-1 elephant grass and closely related cultivars assessed by SRAP markers. Journal of Genetics 2009;88:281-290.

40. Yeh FC, Yang RC, Boyle TBJ, Ye ZH, Mao JX. POPGENE, the user-friendly shareware for population genetic analysis. Canada: Molecular Biology and Biotechnology Centre, University of Alberta; 1997.

Chapter 9

Mineral Nutrient Remobilization During Corolla Senescence in Ethylene Sensitive and -Insensitive Flowers

Michelle L. Jones

Department of Horticulture and Crop Science, The Ohio State University, OARDC, 1680 Madison Avenue, Wooster, OH 44691, USA

ABSTRACT

The flower has a finite lifespan that is controlled largely by its role in sexual reproduction. Once the flower has been pollinated or is no longer receptive to pollination, the petals are programmed to senesce. A majority of the genes that are up-regulated during petal senescence, in both ethylene-sensitive and -insensitive flowers, encode proteins involved in the degradation of nucleic acids, proteins, lipids, fatty acids, and cell wall and membrane components. A smaller subset of these genes has a putative role in remobilizing nutrients, and only a few of these have been studied in detail. During senescence, carbohydrates (primarily sucrose) are transported from petals, and the degradation of macromolecules and organelles also allows the plant to salvage mineral nutrients from the petals before cell death. The remobilization of mineral nutrients from a few species has been investigated and will be reviewed in this article. Ethylene's role in nutrient remobilization is discussed by comparing nutrient changes during the senescence of ethylene-sensitive and -insensitive flowers, and by studies in transgenic petunias (*Petunia* × *hybrida*) that are insensitive to ethylene. Gene expression studies indicate that remobilization is a key feature of senescence, but some senescence-associated genes have different expression in leaves and petals. These gene expression patterns, along with differences in the nutrient content of leaves and petals, suggest that there are differences in the mechanisms of cellular degradation and nutrient transport in vegetative and floral organs. Autophagy may be the mechanism for large-scale degradation that allows for recycling during senescence, but it is unclear if this causes cell death. Future research should focus on autophagy and the regulation of ATG genes by ethylene during both leaf and petal senescence. We must identify the mechanisms by which individual mineral nutrients are transported out of senescing corollas in both ethylene-sensitive and -insensitive species.

INTRODUCTION

The purpose of the angiosperm flower is sexual reproduction, and some plants have colourful and/or fragrant flowers to aid in attracting pollinators. Despite their importance, it is beneficial for the plant to keep the flowers only as long as they are needed (Ashman and Schoen 1994). The maintenance of petals is costly in terms of respiratory energy, nutrients and water loss (Stead 1992; Stead et al. 2006). Once the flower has been pollinated, or the stigma is no longer receptive to pollination, the corolla senesces (Jones 2002). Senescence is the last stage of flower development, during which nutrients are recycled to developing tissues (Bieleski 1995; Verlinden 2003; Chapin and Jones 2007; Trivellini et al. 2011). A genetically controlled senescence programme allows the plant to systematically disassemble cellular organelles and macromolecules, and remobilize nutrients from the petals before cell death. The senescence of many flowers is regulated by the plant hormone ethylene (i.e. ethylene-sensitive flowers), while in other plants the initiation and execution of flower senescence is independent of ethylene (i.e. ethylene-insensitive flowers). A flower's lifespan may be as little as a day or up to several months (van Doorn and Woltering 2008).

While a most intriguing aspect of flower senescence is the signalling between the gynoecium and the corolla, most flower senescence research has focused on the biochemical, molecular and morphological changes occurring in the petals, because it is the longevity of the corolla that determines the post-harvest quality and garden performance of ornamental plants (Jones and Woodson 1999). Senescence begins the developmental transition from anabolism of biomolecules to catabolism of lipids, carbohydrates, proteins and nucleic acids (Jones 2004; Rogers 2006; van Doorn and Woltering 2008; Jones et al. 2009). Common features of petal senescence include losses of cellular organization, increases in membrane permeability and degradation of organelles (Smith et al. 1992; O'Donoghue et al. 2002; Wagstaff et al. 2003). This is accompanied by the increased export of carbohydrates (especially sucrose), macro- and micro-nutrients; decreased nucleic acid and protein content; increased DNA fragmentation; and increased mRNA abundance and enzyme activity of proteases, nucleases and lipases (Bieleski 1995; Stephenson and Rubinstein 1998; Wagstaff et al. 2002; Verlinden 2003; Jones et al. 2005; Langston et al. 2005; Pak and van Doorn 2005; Yamada et al. 2006; Chapin and Jones 2007; Battelli et al. 2011). There is increasing ultrastructural evidence and gene expression data indicating that autophagy occurs during petal senescence, but it is unclear if this is required for cell death (Shibuya et al. 2011). It does seem likely that autophagy is the mechanism by which macromolecules and organelles are disassembled in senescing petals. This allows for the remobilization or translocation of nutrients from the petals to the developing ovary in a pollinated flower, or to other sink tissues when flowers remain unpollinated.

In this article, I will focus on our current understanding of mineral nutrient remobilization during flower petal senescence.

Transcriptome studies using microarrays have identified genes in both ethylene-sensitive and -insensitive flowers that encode catabolic or remobilization proteins, but very little is known about how these genes and their protein products are influenced by pollination, ethylene or environmental stress. The differences in nutrient remobilization during the developmental senescence of unpollinated flowers will be compared with pollinated flowers and with the senescence of short-lived ephemeral flowers. This article will also review critical differences in the senescence of leaves and petals that highlight the importance of conducting thorough comparative studies of these two organs. While many senescence studies have been conducted on cut flowers, an understanding of the biological relevance of nutrient recycling during petal senescence can only come from studies of flowers that are still attached to the plant. Studies of autophagy and transporter proteins (TPs) are increasing our understanding of the mechanisms of cellular degradation and remobilization occurring during senescence, but there is still much to be learned about these processes in flowers.

NUTRIENT REMOBILIZATION DURING PETAL SENESCENCE

Flower longevity is terminated by either corolla wilting or petal abscission. Corolla wilting is accompanied by a gradual decline in petal fresh weight (FW) and a later decrease in petal dry weight (DW). Petals are visibly wilted when epidermal cells lose turgor. The wilting senescence phenotype occurs in both ethylene-sensitive and -insensitive flowers (van Doorn 2001). In some species, corolla wilting is followed by abscission of petals or the entire flower, while other species shed fully turgid corollas. An intermediate wilting phenotype is observed in some flowers that wilt slightly, concurrent with petal abscission. The decrease in petal DW is due to carbon losses from the respiratory climacteric peak and the export of sugars and mineral nutrients (van Doorn and Woltering 2008).

Autolysis during petal senescence occurs first in the interveinal parenchyma, allowing the phloem to remain viable and export nutrients from the senescing petals. Carbohydrates are mainly transported from the petals as sucrose, and [14C]sucrose applied to senescing daylily

(Hemerocallis sp.) tepals is rapidly translocated to developing flower buds through the phloem (Bieleski 1995). Nitrogen (N) is transported out of the petals as soluble amino acids (Eason et al. 2000), and glutamine, hydroxyproline and asparagine are the main transport amino acids identified in the phloem exudates of senescing daylily tepals (Bieleski 1995). Phosphorus (P) is remobilized as the inorganic anion, and many cations, including potassium (K^{+}), magnesium (Mg^{2+}) and calcium (Ca^{2+}), are loaded into the phloem for export during petal senescence (Eason et al. 2000; van Doorn and Woltering 2008).

Changes in petal DW and FW during senescence

In ethylene-insensitive daylily flowers, wilted tepals (petals and sepals) retain only 30–35 % of their maximum DW and <10 % of their FW (Lay Yee et al. 1992; Bieleski 1995). Within 24 h of flower opening, 'Cradle Song' daylily tepals lose 95 % of their sugar content and 65 % of their DW. Only ~8 % of this loss can be attributed to respiration and 30 % is accounted for by decreases in non-sugar components (Bieleski 1995). In contrast, some flowers abscise turgid corollas that have lost little to no FW during flower development. Ethylene-sensitive snapdragon (Antirrhinum majus) flowers continue to increase petal FW until the fully turgid corolla abscises. This is accompanied by a continual increase in petal DW (Goodwin et al. 2003). Flowers like Alstroemeria fall somewhere in between, exhibiting slight wilting of both the petals and sepals followed by abscission. Collier (1997) reported that cut Alstroemeria (Alstroemeria pelegrina) flowers retained only 40 % of their maximum FW, while we have observed less severe wilting in potted Alstroemeria (Alstroemeria hybrid 'Ivana') plants, with flowers still attached to the plant retaining 75–80 % of their FW. Dry weight losses were similar in both the cut flowers and attached flowers, with newly abscised petals retaining 80 % of their maximum DW (Collier 1997; M. L. Jones, unpubl. data).

Macro- And Micro-Nutrient Changes in Petals during the Developmental Senescence of Unpollinated Flowers

Experiments evaluating the macro- and micro-nutrient content of the corolla from anthesis (flower opening) through senescence provide evidence for the remobilization of individual mineral nutrients. A number of experiments evaluating nutrient changes during flower development have been conducted in Petunia × hybrida (Verlinden 2003; Chapin and Jones 2007, 2009). Petunia is often used as a model for ethylene-sensitive flower senescence (Jones et al. 2009). Senescing, unpollinated petunia flowers show visual symptoms of corolla wilting, concomitant with decreases in FW at 6–10 days after anthesis, depending on growing conditions (Verlinden 2003). Decreases in petunia petal DW during senescence are not as extensive as those in daylily, and they retain ~65 % of their maximum DW (Verlinden 2003). The natural senescence of unpollinated 'Mitchell Diploid' (MD) petunia flowers (a.k.a. developmental or age-related senescence) is accompanied by decreases in the N and P content of the corollas (Verlinden 2003; Chapin and Jones 2009). A gradual decline in P throughout development begins well before visual symptoms of wilting or losses in FW. In contrast, the N content of the corolla starts to decline during the later stages of senescence when the petals are wilted and producing ethylene. The P content of unpollinated senescing corollas decreases by 74–75 %, and N decreases by 50–60 % (Verlinden 2003; Chapin and Jones 2009). While a larger percentage of the total P appears to be recovered during the senescence of petunia corollas, this amounts to only 92 µg of P on average per corolla. The N decrease from the corolla averages 390 µg. This loss in P and N represents only 4 and 17 %, respectively, of the total decrease in DW that occurs during the senescence of unpollinated corollas (Chapin and Jones 2009).

It is interesting to note that the amount of P remobilization reported in experiments conducted independently by the Verlinden (West Virginia University) and Jones (The Ohio State University) labs are very similar (74 and 75 % remobilization), while N changes were more variable. Verlinden (2003) also reports a steady decline in the K content of MD petals during the senescence of unpollinated flowers (~40 %), while we have consistently observed that the K content of the corolla increases slightly during development, with final levels in the fully senescent (late-stage senescing) corollas greater than, or the same as, those in corollas at anthesis (Chapin and Jones 2009). In contrast, we have measured a decrease in the K content of ~35 % in senescing corollas from pollinated flowers (Chapin and Jones 2007; see below for more about differences between the senescence of pollinated and unpollinated flowers). These differences may suggest that the remobilization of N and K are influenced more by environmental conditions and the nutritional status of the plant than P. All other macro- and micro-nutrients evaluated either increased in abundance during senescence or were the same when comparing corollas from fully senescent flowers to those at anthesis (Verlinden 2003).

Macro- and Micro-Nutrient Profiles Differ During Pollination-Induced and Developmental Senescence

Pollination accelerates the senescence of many longer-lived flowers (including petunia), and this response is dependent on ethylene signalling (Jones 2008). When petunias are pollinated, the corollas wilt within 48 h (Chapin and Jones 2007). During developmental senescence, wilting of unpollinated petunias occurs at 6–10 days after anthesis. While only the macronutrients N and P (and in some instances K) are remobilized during the senescence of unpollinated flowers (i.e. developmental senescence), additional nutrients are remobilized during pollination-induced senescence in petunia flowers. Nitrogen and P are reduced to a similar extent in both unpollinated and pollinated corollas (Fig. 1A; Verlinden 2003; Chapin and Jones 2007, 2009). In addition, as mentioned previously, the senescent corollas from pollinated flowers have a 35 % reduction in the K content. The macro-nutrient Mg also shows a small, but significant decrease of ~6 %. The micro-nutrients zinc (Zn) and copper (Cu) decrease during pollination-induced senescence, with the largest decreases in Cu content (54 % reduction).

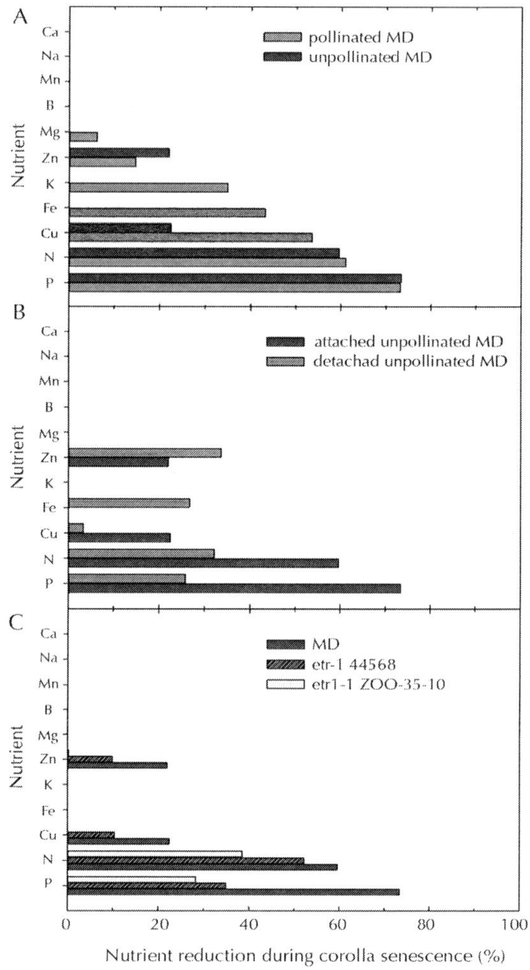

Figure 1: Summary of the percent reduction in macro-nutrients and micro-nutrients during Petunia × hybrida corolla senescence. The total nutrient content of corollas at anthesis was compared with nutrient content at the advanced stage of senescence (fully wilted) as an indication of nutrient re-mobilization. (A) Nutrient remobilization in MD petunia corollas during pollination-induced senescence and during the developmental senescence of unpollinated corollas. (B) Nutrient remobilization in MD corollas during the senescence of unpollinated flowers attached to the plant or in detached, cut flowers. (C) Nutrient remobilization in MD wild-type corollas and transgenic

corollas (lines Z00-35-10 and 44568) with reduced sensitivity to ethylene due to the expression of the mutated ethylene receptor (35S::Atetr1-1). Line 44568 etr1-1 is less sensitive to ethylene than line Z00-35-10 etr1-1.

In pollinated petunias (detached flowers), N increases in the ovary were found to occur concomitantly with decreases in petal N content (Shibuya et al. 2013), suggesting that the nutrients that are remobilized from pollinated senescing corollas are transported to the developing ovary. Unfortunately, N was the only nutrient measured in this study. The post-pollination response in orchids is also associated with increased ethylene production and accelerated wilting of the perianth (*calyx plus corolla*). Pollination results in an increase in the FW and DW of the gynostemium and the ovary, and a decrease in the FW and DW of the perianth (Hew et al. 1989). These changes have been associated with the transport of N and phosphate from the perianth to the ovary in *Cymbidium* and *Arachnis* orchids (Harrison and Arditti 1976; Hew et al. 1989). The movement of these nutrients is dictated by sink activities. When radiolabelled P is applied to the labella (lips and median petals) of unpollinated Cymbidium flowers, 32P is detected throughout the flower, with the lowest levels in the perianth. Pollination redirects this transport, and the largest accumulation of ^{32}P is detected in the *gynostemia* (column) and the ovaries (Harrison and Arditti 1976). The N and P gains in the ovary of Arachnis orchids are much greater than the losses from the perianth, suggesting that ovule development also requires the additional remobilization of nutrients from sources outside the flower (Hew et al. 1989).

It has been proposed that pollination-induced ethylene production in the orchid flower regulates the changes in sink activity of the ovary that drive nutrient transport (Hew et al. 1989). Treating carnation flowers with exogenous ethylene mimics pollination, accelerating the wilting of the petals and increasing the growth of the ovary (Nichols 1976). In this experiment, cut flowers were treated with 1 $\mu l\ L^{-1}$ ethylene for 1 day. After 4 days, the DW of the petals decreases by 29% and this is accompanied by a decrease in N (43 %), P (53 %) and K (27 %). Ovary DW increases and the N, P and K content of the ovary also increases (Nichols 1976). These experiments provide support for ethylene's role in remobilization of nutrients from the petals to the developing ovary.

Transporting nutrients within the plant requires energy, so the specific nutrient and the amount of each nutrient that is remobilized must result in a net advantage to plant growth and reproductive development. The largest remobilization from corollas occurs with N and P, which are the most likely to limit plant growth, and would therefore be the most advantageous for the plant to recycle. It takes an additional investment of energy and resources to remobilize nutrients outside the flower (when flowers remain unpollinated) versus within the flower (following pollination or ethylene treatment). This additional 'cost' of recycling nutrients from unpollinated flowers may explain the differences in the remobilization profiles observed during developmental and pollination-induced senescence in longer-lived flowers. Additional experiments with ^{32}P are needed to track the movement of nutrients from the petals to the developing ovary, and to determine when nutrients are remobilized to alternative sinks.

Nutrient Remobilization during the Senescence of Ephemeral Flowers

Perhaps more surprising is the fact that the very rapid senescence program of ephemeral flowers also involves a nutrient remobilization phase. The changes in the nutrient profiles of senescing Hibiscus rosa-sinensis (*hibiscus*) are very similar to the pollination-induced changes in petunia corollas (Chapin and Jones 2007; Trivellini et al. 2011). Hibiscus is an ethylene-sensitive, ephemeral flower that opens and wilts within 1 day. This very rapid senescence program is accompanied by decreases in P, N, Cu, K, Zn and Mg, the exact same nutrients that are remobilized during the accelerated senescence of pollinated petunia corollas. In this experiment, hibiscus plants were grown in the greenhouse. There was no mention of hand pollination, and the greenhouse environment would reduce the chance of pollination by birds or insects. The lack of an increase in the DW of ovaries from the open to senescing flower stages also supports the assumption that these flowers were not pollinated. Similarly, the senescence of ethylene-sensitive Ipomoea purpurea (morning glory) flowers is accompanied by a 75 % reduction in the P content of the petals, and a 65, 55 and 34 % reduction in the amount of K, Mg and Ca, respectively (*Winkenbach* 1970). The senescence of ephemeral flowers is generally not accelerated by pollination. Nutrient remobilization studies indicate that the senescence of these flowers may be more similar to pollination-induced senescence than developmental senescence in their longer-lived counterparts (Winkenbach 1970; Trivellini et al.

2011). In Japanese morning glories, petal senescence starts before the flowers are even open (Winkenbach 1970). This early initiation of senescence would be needed to allow for cellular degradation and nutrient remobilization in such short-lived flowers.

Senescence of Cut Flowers versus Flowers that are attached to the Plant

Many studies of flower senescence have been conducted in cut flowers that have been harvested from the plant and held in water or a vase solution (Wagstaff et al. 2003; Breeze et al. 2004; Pak and van Doorn 2005; Hoeberichts et al. 2007; Battelli et al. 2011). While excellent studies of senescence have been conducted in these systems, we must interpret these results with caution when it comes to the nutrient remobilization phase of senescence. In a cut stem with a single flower, like standard carnations (*Dianthus caryophyllus*), nutrient remobilization from petals of unpollinated flowers may be reduced compared with flowers that are still attached to the plant, if nutrient translocation is driven by sinks outside of that flowering stem. In contrast, inflorescence or spike-type cut flowers, which have multiple flowers at different developmental stages, may provide an excellent model for studies of remobilization from senescing to developing flowers. Removal of the lower three open flowers from a *Gladiolus* spike reduces the total DW of newly opened florets, indicating that the older flowers on the spike are an important source of nutrients for the newly developing flowers (Waithaka et al. 2001).

To investigate nutrient remobilization in a cut flower system, detached, single petunia flowers were placed in a vial of distilled water and allowed to senesce. The nutrient content of corollas at anthesis (on the day of flower opening from a flower just harvested from the plant) was compared with the nutrient content of corollas from detached flowers that had been allowed to senesce off the plant (Fig. 1B; M. L. Jones, unpubl. data). In the detached flowers, the N and P content of the corollas decreases during senescence, but remobilization is reduced compared with corollas from attached flowers that senesced on the plant. The reduction in N is only 32 % and P is 26 %, compared with 60 and 74 % in attached flowers. The only other difference between attached and detached flowers is a decrease in the iron (Fe) content (27 %) of senescing detached corollas that is not detected in attached corollas. The senescence of detached daylily flowers is visually similar to the senescence of attached flowers, but there are major differences in tepal composition. Senescing tepals from detached flowers have a higher FW, DW, sugar and amino acid content than tepals from attached flowers (Bieleski 1995). These studies show that nutrient remobilization from the petals is reduced when flowers are detached from the plant during senescence. Further studies are needed to determine the signals that regulate nutrient remobilization, but it is clear that studies of developmental senescence should include flowers that are still attached to the plant.

Ethylene Regulation of Nutrient Remobilization during Petal Senescence

Macro- and micro-nutrient changes have also been investigated in transgenic petunias that have reduced sensitivity to ethylene due to ectopic expression of the mutated ethylene receptor etr1-1 (35S::Atetr1-1; Wilkinson et al. 1997). These plants serve as an excellent model system for investigating the role of ethylene in nutrient remobilization (Chapin and Jones 2009). Two different transgenic lines, Z00-35-10 and 44568 (obtained from David Clark, University of Florida), were used for these studies. Both lines have extremely reduced responsiveness to ethylene, but line 44568 is less sensitive than Z00-35-10.

In etr1-1 flowers, the N content is reduced by 52 and 38 % during the senescence of unpollinated corollas in the Z00-35-10 and 44568 lines, respectively (Fig. 1C; Chapin and Jones 2009). This is compared with a 60 % reduction in the MD flowers. While ethylene signalling may influence the transport of N out of the senescing petals, the influence of ethylene on P remobilization during senescence is much more striking. While 73 % of the P is remobilized from MD corollas, only 35 and 28 % of the P is remobilized from the etr1-1 Z00-35-10 and 44568 lines, respectively. The final seed weight of etr1-1 petunias is also significantly different from that of MD flowers, which may be in part due to reduced N and P remobilization to the ovary (Clevenger et al. 2004).

The Involvement of TPS in Nutrient Remobilization within the Flower

Transporter proteins are critically involved in the process of nutrient remobilization. Large-scale expression profiling in leaves has identified TPs from many different classes that are up-regulated during the senescence of Arabidopsis (*Arabidopsis thaliana*) leaves (van der Graaff et al. 2006), but fewer TPs have been identified in petals. Even fewer have been characterized to any extent to investigate their role in petal senescence, or to determine how they are regulated by ethylene. Microarray studies in Alstroemeria and carnation petals have identified senescence up-regulated genes that are classified as transporters (Breeze et al. 2004; Hoeberichts et al. 2007). In carnation petals, a metal–nicotianamine (NA) transporter (Hoeberichts et al. 2007) is up-regulated during petal senescence. These NA transporters belong to the oligopeptide transporter family and have been implicated in metal transport in plants. Also called yellow stripe-like (YSL) proteins in rice, they are involved in metal detoxification and specifically transport iron and manganese (Mn)–NA chelates in the phloem (Ishimaru et al. 2010; Sasaki et al. 2011). The role of YSL proteins in petals warrants further investigation, and may shed some light on the role of metal transport in and out of the flower during development and senescence.

Five high-affinity phosphate transporter genes have been identified in petunia (Wegmuller et al. 2008; Chapin and Jones 2009). The majority of these are expressed in roots and believed to be involved in Pi uptake. PhPT1 transcripts are detectable at low levels in non-senescing petals and increase during senescence (Chapin and Jones 2009). The highest transcript abundance correlates with the large decreases in P content that accompany corolla senescence. *PhPT1* transcript abundance is reduced in both Z00-35-10 and 44568 $etr1^{-1}$ petunias, supporting the aforementioned conclusion that P remobilization involves ethylene signalling. Treating flowers with a very low concentration of ethylene (0.1 μl L^{-1}) for only 2 h results in up-regulation of *PhPT1* transcripts (Chapin and Jones 2009). Treatment with cycloheximide does not prevent this induction, indicating that de novo protein synthesis is not required for this response, and therefore PhPT1 is an ethylene primary response gene. This is unusual for a gene involved in the nutrient remobilization phase of senescence. Most of the senescence-associated genes, including *SAG12* in Arabidopsis, are up-regulated by ethylene as a secondary response (Weaver et al. 1998). This can explain why P levels start to decrease in the petals earlier than N. The primary regulation of *PhPT1* by ethylene may also allow the plant to use the P in the corolla as a reserve during P starvation or other stresses that induce senescence.

A putative nitrate transporter (*DAFSAG7*) is up-regulated during the senescence of *Narcissus pseudonarcissus* (daffodil) tepals. Transcripts are not detected in the perianth of newly open flowers, but are first detected at 3 days after opening, just prior to tepal discolouration. Transcript abundance increases during perianth senescence, supporting a role for this nitrate transporter in the nutrient remobilization phase of flower senescence (Hunter et al. 2002). A putative peptide transporter (*WC11*) is up-regulated in both *Erysimum linifolium* (wallflower) petals and leaves (though to a greater extent in petals) (Price et al. 2008). WC11 has a much higher basal level in non-senescing leaves. It is up-regulated early in the senescence programme and decreases by the time that leaves are visually yellowing and *SAG12* (often used as a marker for senescence) is up-regulated. The WC11 transcripts are induced to a greater extent early in petals, well before visual senescence, and remain high throughout senescence (Price et al. 2008). The closest Arabidopsis homologue to *WC11* is At1g32450, which encodes a xylem nitrate-loading transporter that is involved in nitrate allocation to roots and responses to drought, salinity and cadmium stress (Chen et al. 2012). This expression pattern does not match what we would have expected based on the late N export patterns from petunia corollas, but may suggest that this transporter also has a role in N transport into young sink tissues (either leaves or petals).

AUTOPHAGY

Recent research has suggested that autophagy is the mechanism for the degradation of macromolecules during the senescence of flower petals (Yamada et al. 2009). Autophagy is a type of programmed cell death first identified in animal cells that is characterized by the appearance of *autophagosomes*, vacuole-like vesicles involved in degradation of the cytosol (van Doorn 2011). This process is also defined by the activation of autophagy or ATG genes. Homologues to the Arabidopsis autophagy-related proteins *ATG4b* and *ATG8a* have been identified in *Ipomoea nil* (Japanese morning glory) petals (Yamada et al. 2009; Shibuya et al. 2011). Morning glories are ethylene-sensitive, ephemeral flowers that open and senesce in 1 day, and the senescence program in mesophyll cells is initiated even prior to flower opening. Petal senescence is accompanied by an increase in transcript abundance for both In*ATG4b* and In*ATG8a*. Ethylene treatments that accelerate flower senescence up-regulate InATG4b but not InATG8a expression (Yamada et al. 2009). The presence of *autophagosome*-like vesicles has also been demonstrated in the cytoplasm and the vacuole (Matile and Winkenbach 1971; Matile 1997). *Monodansylcadaverine,* which has been shown to stain *autophagosomes,* has identified the appearance of autophagic structures during petal senescence in both morning glory and petunia (Shibuya et al. 2009, 2013). In petunia flowers, pollination-induced and developmental senescence is accompanied by an increase in transcript abundance for *PhATG8* homologues (*ATG8a, ATG8b, ATG8c and ATG8d*) (Shibuya et al. 2013). Additional research is needed to determine the differential regulation of ATG genes by ethylene. Functional analysis of the ATG genes in petunia using *RNAi* or virus-induced gene silencing should provide more evidence for or against the role of autophagy in developmental and pollination-induced petal senescence.

DIFFERENCES BETWEEN LEAF AND PETAL SENESCENCE

Few studies of leaf and petal senescence have been conducted in the same species, as the ideal model systems for petal senescence, including *petunia,* do not have distinct and predictable leaf senescence like Arabidopsis. It is also difficult to identify physiological or morphological characteristics that can be used as developmental benchmarks to ensure that equivalent senescence stages are being compared between leaves and petals. Comparative analyses of gene expression in *Arabidopsis* (Wagstaff et al. 2009) and wallflower (Price et al. 2008), a relative of *Arabidopsis* that has large ornamental flowers, have identified key metabolic pathways that are shared between senescing leaves and petals, as well as genes that are up-regulated only in petals. These studies have identified senescence pathways that are tissue specific and which validate the need to study senescence in both petals and leaves. Genes associated with autophagy, degradation and remobilization are up-regulated in both leaves and petals of Arabidopsis and wallflower, but differences in the expression profiles of specific genes are apparent (Price et al. 2008; Wagstaff et al. 2009). Within the *APG8/ATG8* subfamily in Arabidopsis, members *APG8a* and *APG8b* are up-regulated during both petal and leaf senescence. In contrast, *APG8f* and APG8i are up-regulated during petal senescence, but earlier in leaf development (at 16 days after emergence). *APG8e* shows little change during leaf development and senescence, but is up-regulated during petal senescence (Wagstaff et al. 2009). These findings suggest that differences in cellular constituents and nutrient content have led to the differences in autophagy that allow for the specific mechanisms of degradation and nutrient remobilization in senescing leaves and petals. In wallflower, the identification of genes encoding proteins involved in degradation and remobilization (e.g. a copper chaperone, lipid transfer protein, thiol protease and AAA-type ATPase family protein), which are up-regulated during petal senescence and not during leaf senescence, provides additional evidence for differences in the remobilization phase of senescence in petals versus leaves (Price et al. 2008).

Studies in Arabidopsis indicate that 80 % or greater of the N, P and K content is remobilized from senescing leaves. In addition to the Cu, Zn and Mg that are remobilized in senescing (pollinated) petunia and hibiscus petals, levels of chromium (Cr), Fe and molybdenum (Mo) also drop by >40 % during Arabidopsis leaf senescence. Smaller reductions (~15 %) are observed in the cobalt (Co), nickel (Ni), Mn, sodium (Na) and Ca content (Himelblau and Amasino 2001). In the comparative study of leaf and petal senescence in wallflower, the extent of protein degradation was found to be much greater during the senescence of leaves. The senescent leaves (stage 7) retain only 5 % of maximum protein levels compared with 65 % in senescent petals (stage 5) (Price et al. 2008). This decrease in protein content is concomitant with chlorophyll degradation, and is not surprising considering that the majority of the remobilization occurring in leaves is due to the degradation of chloroplast proteins (Thomas and Donnison 2000).

Petals are fairly nutrient poor compared with leaves, and these experiments indicate that leaves remobilize a greater total amount and larger number of individual nutrients during senescence (Himelblau and Amasino 2001; Verlinden 2003; Chapin and Jones 2007, 2009). When comparing young, newly expanded green petunia leaves to petunia corollas on the day of anthesis (Fig. 2), the largest difference is the N content. Leaves contain 63 mg N per g DW, while petals contain only 18 mg N per g DW. That is over 300 % more N in leaves than in petals. The difference between the P and K content of leaves and petals is much smaller. Of the other macro-nutrients, the largest difference between leaves and petals is seen with Ca. Among the micro-nutrients, the greatest differences between leaves and petals are in B, Fe and Mn content. The nutrient content of etr1-1 petals is similar to that of MD petals, indicating that ethylene signalling is not required for the accumulation of nutrients in non-senescing, developing corollas. The nutrient content of young leaves is also similar, with the exception of N, which is consistently higher in etr1-1 leaves, and B, which is higher in MD leaves (Fig. 2).

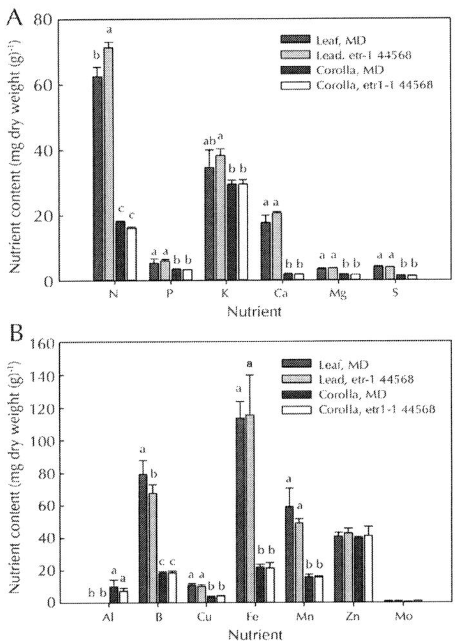

Figure 2: Nutrient content of leaves and corollas from Petunia × hybrida. The (A) macro-nutrient and (B) micro-nutrient content of corollas from flowers at anthesis was compared with the nutrient content of young, newly expanded green leaves. Plants included wild-type MD and transgenic etr1-1 (44568) petunias with reduced sensitivity to ethylene.

Interestingly, the micro-nutrients that are remobilized from pollinated petals, Cu and Zn, are also nutrients that have the same or similar concentrations in leaves and petals. If specific nutrients are much higher in the leaves than in the petals, plants may invest in recycling them only from the more nutrient-rich tissues. If specific nutrients are needed as cofactors or activators for senescence-related enzymes, this may explain why these nutrients continue to increase in abundance or remain at similar levels, rather than being remobilized out of the petals during senescence. For example, Fe is at a much lower concentration in petals than leaves, and it is retained in the petals during flower senescence. Iron is a cofactor for the enzyme ACC oxidase, which catalyses the autocatalytic burst of ethylene that is synthesized by the petals during senescence (Zhang et al. 1995). The petals may need to retain Fe during senescence to allow for ethylene production.

CONCLUSIONS

Flowers may abscise fully turgid corollas or petals may be severely wilted before they are eventually shed. There are also intermediate wilting phenotypes that fall in between these two extremes. The wilting of flowers is accompanied by remobilization of mineral nutrients (primarily N and P) from the petals. This nutrient remobilization has been demonstrated in ethylene-sensitive flowers (e.g. petunia, hibiscus, morning glory and orchids) and ethylene-insensitive flowers (e.g. daylily). Ethylene appears to regulate some components of the remobilization pathways in ethylene-sensitive flowers, as is seen with P transport out of senescing petunia petals and expression of the phosphate transporter PhPT1 (Chapin and Jones 2009). Alstroemeria flowers, which senesce and abscise independently of ethylene action, show an intermediate wilting phenotype. While gene expression changes indicate up-regulation of proteases and other genes encoding proteins involved in cellular degradation (Wagstaff et al. 2002; Breeze et al. 2004), it is not clear how much of the senescence programme is completed in these flowers before the petals and sepals are shed. It can be assumed that senescence is not occurring in petals from flowers (e.g. digitalis and snapdragon) that abscise fully turgid corollas, but there are no gene expressions or nutrient remobilization studies to support or refute this claim. Nutrient changes must be investigated in abscission type, and more importantly, in intermediate wilting type flowers, to provide us with a more complete understanding of the role that nutrient remobilization from the petals plays in the terminal phases of flower development.

Increasing evidence suggests that autophagy is involved in petal senescence, but studies are needed to characterize the expression of individual ATG genes in petals from multiple species, and to determine if this process is regulated by ethylene. It is not clear if autophagy is the cause of cell death, or if it is a mechanism that allows for efficient remobilization by delaying cell death until nutrients can be recycled. While no one seems to disagree that nutrient recycling is the central function of senescence in both leaves and petals, it is less clear if and when mineral nutrient remobilization from petals will have an important impact on the overall fitness or reproductive success of the plant. I have postulated that salvaging N or P, nutrients that are most likely to limit plant growth, from the petals may have a positive impact on the overall nutrient status of the plant. While this may be true, it may be under nutrient starvation conditions where recycling mineral nutrients from the petals is most important. The nutrients that can be remobilized from the corolla to the ovary and developing seeds may be increasingly important under these conditions and may allow for successful reproduction under environmental stress. This is an area of flower senescence that has received little attention and must be addressed to truly understand the biological implications of petal senescence.

Comparative transcriptome analyses between flowers with different senescence syndromes have been limited by the lack of sequence data in all but a few model species. The increased accessibility and decreased cost of next-generation sequencing should result in a large increase in sequence availability, and allow for more comprehensive investigations of the petal senescence transcriptomes in ethylene-sensitive and - insensitive species. These experiments will also allow us to identify the extent of the senescence programme that occurs in species like Alstroemeria, which show limited wilting before the petals are shed. These studies, in combination with proteomic (Bai et al. 2010) and metabolomic investigations, will help us to further our understanding of the biological relevance of nutrient remobilization during petal senescence, and allow us to determine how petal senescence is regulated by ethylene, during development and under stress.

REFERENCES

1. Ashman TL, Schoen DJ. How long should flowers live? Nature 1994;371:788-791. doi:10.1038/371788a0.
2. Bai S, Willard B, Chapin LJ, Kinter MT, Francis D, Stead AD, Jones ML. Proteomic analysis of pollination-induced corolla senescence in petunia. Journal of Experimental Botany 2010;61:1089-1109. doi:10.1093/jxb/erp373.
3. Battelli R, Lombardi L, Rogers HJ, Picciarelli P, Lorenzi R, Ceccarelli N. Changes in the ultrastructure, protease and caspase-like activities during flower senescence in Lilium longiflorum. Plant Science 2011;180:716-725. doi:10.1016/j.plantsci.2011.01.024.
4. Bieleski RL. Onset of phloem export from senescent petals of daylily. Plant Physiology 1995;109:557-565.
5. Breeze E, Wagstaff C, Harrison E, Bramke I, Rogers H, Stead A, Thomas B, Buchanan-Wollaston V. Gene expression patterns to define stages of post-harvest senescence in Alstroemeria petals. Plant Biotechnology Journal 2004;2:155-168. doi:10.1111/j.1467-7652.2004.00059.x.
6. Chapin LJ, Jones M. Nutrient remobilization during pollination-induced corolla senescence in Petunia. Acta Horticulturae 2007;755:181-190.
7. Chapin LJ, Jones M. Ethylene regulates phosphorus remobilization and expression of a phosphate transporter (PhPT1) during petunia corolla senescence. Journal of Experimental Botany 2009;60:2179-2190. doi:10.1093/jxb/erp092.
8. Chen CZ, Lv XF, Li JY, Yi HY, Gong GM. Arabidopsis NRT5.1 is another essential component in the regulation of nitrate reallocation and stress tolerance. Plant Physiology 2012;159:1582-1590. doi:10.1104/pp.112.199257.
9. Clevenger DJ, Barrett JE, Klee HJ, Clark DG. Factors affecting seed production in transgenic ethylene-insensitive petunias. Journal of

the American Society for Horticultural Science 2004;129:401-406.

10. Collier DE. Changes in respiration, protein and carbohydrates of tulip tepals and Alstroemeria petals during development. Journal of Plant Physiology 1997;150:446-451. doi:10.1016/S0176-1617(97)80096-X.

11. Eason JR, Johnston JW, de Vre L, Sinclair BK, King GA. Amino acid metabolism in senescing Sandersonia aurantiaca flowers: cloning and characterization of asparagine synthetase and glutamine synthetase cDNAs. Australian Journal of Plant Physiology 2000;27:389-396.

12. Goodwin SM, Kolosova N, Kish CM, Wood KV, Dudareva N, Jenks MA. Cuticle characteristics and volatile emissions of petals in Antirrhinum majus. Physiologia Plantarum 2003;117:435-443. doi:10.1034/j.1399-3054.2003.00047.x.

13. Harrison CR, Arditti J. Post-pollination phenomena in orchid flowers. VII. Phosphate movement among floral segments. American Journal of Botany 1976;63:911-918. doi:10.2307/2441748.

14. Hew CS, Tan SC, Chin TY, Ong TK. Influence of ethylene on enzyme activities and mobilization of materials in pollinated Arachnis orchid flowers. Journal of Plant Growth Regulation 1989;8:121-130. doi:10.1007/BF02025279.

15. Himelblau E, Amasino RM. Nutrients mobilized from leaves of Arabidopsis thaliana during leaf senescence. Journal of Plant Physiology 2001;158:1317-1323. doi:10.1078/0176-1617-00608.

16. Hoeberichts FA, van Doorn WG, Vorst O, Hall RD, van Wordragen MF. Sucrose prevents up-regulation of senescence-associated genes in carnation petals. Journal of Experimental Botany 2007;58:2873-2885. doi:10.1093/jxb/erm076.

17. Hunter DA, Steele BC, Reid MS. Identification of genes associated with perianth senescence in daffodil (Narcissus pseudonarcissus L. 'Dutch Master'). Plant Science 2002;163:13-21. doi:10.1016/S0168-9452(02)00068-7.

18. Ishimaru Y, Masuda H, Bashir K, Inoue H, Tsukamoto T, Takahashi M, Nakanishi H, Aoki N, Hirose T, Ohsugi R, Nishizawa NK.

Rice metal-nicotianamine transporter, OsYSL2, is required for the long-distance transport of iron and manganese. The Plant Journal 2010;62:379-390. doi:10.1111/j.1365-313X.2010.04158.x.

19. Jones ML. Ethylene responsiveness in carnation styles is associated with stigma receptivity. Sexual Plant Reproduction 2002;15:107-112. doi:10.1007/s00497-002-0146-4.
20. CrossRefWeb of Science
21. Jones ML. Changes in gene expression during senescence. In: Nooden L, editor. Plant cell death processes. San Diego, CA: Elsevier Science; 2004. p. 51-72.
22. Jones ML. Ethylene signaling is required for pollination-accelerated corolla senescence in petunias. Plant Science 2008;175:190-196. doi:10.1016/j.plantsci.2008.03.011.
23. Jones ML, Woodson WR. Differential expression of three members of the ACC synthase gene family in carnation. Plant Physiology 1999;119:755-764. doi:10.1104/pp.119.2.755.
24. Jones ML, Chaffin GS, Eason JR, Clark DG. Ethylene-sensitivity regulates proteolytic activity and cysteine protease gene expression in petunia corollas. Journal of Experimental Botany 2005;56:2733-2744. doi:10.1093/jxb/eri266.
25. Jones ML, Stead AD, Clark DG. Petunia flower senescence. In: Gerats T, Stommer J, editors. Petunia: a model system for comparative research. New York: Springer; 2009. p. 301-324.
26. Langston BJ, Bai S, Jones ML. Increases in DNA fragmentation and induction of a senescence-specific nuclease are delayed during corolla senescence in ethylene-insensitive (etr1-1) transgenic petunias. Journal of Experimental Botany 2005;56:15-23. doi:10.1093/jxb/eri124.
27. Lay Yee M, Stead AD, Reid MS. Flower senescence in daylily (Hemerocallis). Physiologia Plantarum 1992;86:308-314. doi:10.1034/j.1399-3054.1992.860218.x.
28. Matile P. The vacuole and cell senescence. Advances in Botanical Research 1997;25:87-111. doi:10.1016/S0065-2296(08)60149-4.
29. Matile P, Winkenbach F. Function of lysosomes and lysosomal enzymes in the senescing corolla of the morning glory (Ipomoea

purpurea). Journal of Experimental Botany 1971;22:759-761. doi:10.1093/jxb/22.4.759.

30. Nichols R. Cell enlargement and sugar accumulation in the gynoecium of the glasshouse carnation (Dianthus caryophyllus L.) induced by ethylene. Planta 1976;130:47-52. doi:10.1007/BF00390843.
31. O'Donoghue EM, Somerfield SD, Heyes JA. Organisation of cell walls in Sandersonia aurantiaca floral tissue. Journal of Experimental Botany 2002;53:513-523. doi:10.1093/jexbot/53.368.513.
32. Pak C, van Doorn WG. Delay of Iris flower senescence by protease inhibitors. New Phytologist 2005;165:473-480. doi:10.1111/j.1469-8137.2004.01226.x.
33. Price AM, Orellana ADF, Salleh FM, Stevens R, Acock R, Buchanan-Wollaston V, Stead AD, Rogers HJ. A comparison of leaf and petal senescence in wallflower reveals common and distinct patterns of gene expression and physiology. Plant Physiology 2008;147:1898-1912. doi:10.1104/pp.108.120402.
34. Rogers HJ. Programmed cell death in floral organs: how and why do flowers die? Annals of Botany 2006;97:309-315. doi:10.1093/aob/mcj051.
35. Sasaki A, Yamaji N, Xia JX, Ma JF. OsYSL6 is involved in the detoxification of excess manganese in rice. Plant Physiology 2011;157:1832-1840. doi:10.1104/pp.111.186031.
36. Shibuya K, Yamada T, Suzuki T, Shimizu K, Ichimura K. InPSR26, a putative membrane protein, regulates programmed cell death during petal senescence in Japanese morning glory. Plant Physiology 2009;149:816-824. doi:10.1104/pp.108.127415.
37. Shibuya K, Shimizu K, Yamada T, Ichimura K. Expression of autophagy-associated ATG8 genes during petal senescence in Japanese morning glory. Journal of the Japanese Society for Horticultural Science 2011;80:89-95. doi:10.2503/jjshs1.80.89.
38. Shibuya K, Niki T, Ichimura K. Pollination induces autophagy in petunia petals via ethylene. Journal of Experimental Botany 2013;64:1111-1120. doi:10.1093/jxb/ers395.
39. Smith MT, Saks Y, van Staden J. Ultrastructural changes in the

petals of senescing flowers of Dianthus caryophyllus L. Annals of Botany 1992;69:277-285.

40. Stead AD. Pollination-induced flower senescence: a review. Plant Growth Regulation 1992;11:13-20. doi:10.1007/BF00024427.
41. Stead AD, van Doorn WG, Jones ML, Wagstaff C. Flower senescence: fundamental and applied aspects. In: Ainsworth C, editor. Flowering and its manipulation. Oxford, UK: Blackwell; 2006. p. 261-296. Annual Plant Reviews, Vol. 20.
42. Stephenson P, Rubinstein B. Characterization of proteolytic activity during senescence in daylily. Physiologia Plantarum 1998;104:463-473. doi:10.1034/j.1399-3054.1998.1040323.x.
43. Thomas H, Donnison I. Back from the brink: plant senescence and its reversibility. In: Bryant J, Garland JM, Hughes SG, editors. Programmed cell death in animals and plants. Oxford, UK: BIOS Scientific; 2000. p. 149-162.
44. Trivellini A, Ferrante A, Vernieri P, Carmassi G, Serra G. Spatial and temporal distribution of mineral nutrients and sugars throughout the lifespan of Hibiscus rosa-sinensis L. flower. Central European Journal of Biology 2011;6:365-375. doi:10.2478/s11535-011-0025-9.
45. van der Graaff E, Schwacke R, Schneider A, Desimone M, Flugge U-I, Kunze R. Transcription analysis of Arabidopsis membrane transporters and hormone pathways during developmental and induced leaf senescence. Plant Physiology 2006;141:776-792. doi:10.1104/pp.106.079293.
46. van Doorn WG. Categories of petal senescence and abscission: a re-evaluation. Annals of Botany 2001;87:447-456. doi:10.1006/anbo.2000.1357.
47. van Doorn WG. Classes of programmed cell death in plants, compared to those in animals. Journal of Experimental Botany 2011;62:4749-4761. doi:10.1093/jxb/err196.
48. van Doorn WG, Woltering EJ. Physiology and molecular biology of petal senescence. Journal of Experimental Botany 2008;59:453-480. doi:10.1093/jxb/erm356.
49. Abstract/FREE Full Text
50. Verlinden S. Changes in the mineral nutrient concentrations

in petunia corollas during development and senescence. HortScience 2003;38:71-74.

51. ↵ Wagstaff C, Leverentz MK, Griffiths G, Thomas B, Chanasut U, Stead AD, Rogers HJ. Protein degradation during senescence of Alstroemeria petals. Journal of Experimental Botany 2002;53:233-240. doi:10.1093/jexbot/53.367.233.
52. Wagstaff C, Malcolm P, Rafiq A, Leverentz M, Griffiths G, Thomas B, Stead A, Rogers H. Programmed cell death (PCD) processes begin extremely early in Alstroemeria petal senescence. New Phytologist 2003;160:49-59. doi:10.1046/j.1469-8137.2003.00853.x.
53. Wagstaff C, Yang TJW, Stead AD, Buchanan-Wollaston V, Roberts JA. A molecular and structural characterization of senescing Arabidopsis siliques and comparison of transcriptional profiles with senescing petals and leaves. The Plant Journal 2009;57:690-705. doi:10.1111/j.1365-313X.2008.03722.x.
54. Waithaka K, Dodge LL, Reid MS. Carbohydrate traffic during opening of gladiolus florets. Journal of Horticultural Science and Biotechnology 2001;76:120-124.
55. Weaver LM, Gan S, Quirino B, Amasino RM. A comparison of the expression patterns of several senescence-associated genes in response to stress and hormone treatment. Plant Molecular Biology 1998;37:455-469. doi:10.1023/A:1005934428906.
56. Wegmuller S, Svistoonoff S, Reinhardt D, Stuurman J, Amrhein N, Bucher M. A transgenic dTph1 insertional mutagenesis system for forward genetics in mycorrhizal phosphate transport of Petunia. The Plant Journal 2008;54:1115-1127. doi:10.1111/j.1365-313X.2008.03474.x.
57. Wilkinson JQ, Lanahan MB, Clark DG, Bleecker AB, Chang C, Meyerowitz EM, Klee HJ. A dominant mutant receptor from Arabidopsis confers ethylene insensitivity in heterologous plants. Nature Biotechnology 1997;15:444-447. doi:10.1038/nbt0597-444.
58. Winkenbach F. Zum stoffwechsel der aufblühenden und welkenden korolle der prunkwinde Ipomoea purpurea: I. Beziehungen zwischen gestaltwandel, stofftransport, atmung und intertaseaktivität. Berichte der Schweizerischen Botanischen Gesellschaft 1970;80:374-390.

59. Yamada T, Takatsu Y, Kasumi M, Ichimura K, van Doorn WG. Nuclear fragmentation and DNA degradation during programmed cell death in petals of morning glory (Ipomoea nil). Planta 2006;224:1279-1290. doi:10.1007/s00425-006-0307-z.

60. Yamada T, Ichimura K, Kanekatsu M, van Doorn WG. Homologs of genes associated with programmed cell death in animal cells are differentially expressed during senescence of Ipomoea nil petals. Plant and Cell Physiology 2009;50:610-625. doi:10.1093/pcp/pcp019.

61. Zhang Z, Schofield CJ, Baldwin JE, Thomas P, John P. Expression, purification and characterization of 1-aminocyclopropane-1-carboxylate oxidase from tomato in Escherichia coli. Biochemical Journal 1995;307:77-85.

Chapter 10

Characterization of TM8, a MADS-box Gene Expressed in Tomato Flowers

Margherita Daminato[1], Simona Masiero[2], Francesca Resentini[2], Alessandro Lovisetto[1], and Giorgio Casadoro[1, 3]

[1]Department of Biology, University of Padua, Via G. Colombo, 3, 35131 Padua, Italy

[2]Department of Bioscience, University of Milan, Via Celoria, 26, 20133 Milan, Italy

[3]Botanical Garden, University of Padua, Via Orto Botanico, 15, 35123 Padua, Italy

ABSTRACT

Background

The identity of flower organs is specified by various MIKC MADS-box transcription factors which act in a combinatorial manner. *TM8* is a MADS-box gene that was isolated from the floral meristem of a tomato mutant more than twenty years ago, but is still poorly known from a functional point of view in spite of being present in both Angiosperms and Gymnosperms, with some species harbouring more than one copy of the gene. This study reports a characterization of *TM8* that was carried out in transgenic tomato plants with altered expression of the gene.

Results

Tomato plants over-expressing either *TM8* or a chimeric repressor form of the gene (*TM8:SRDX*) were prepared. In the *TM8* up-regulated plants it was possible to observe anomalous stamens with poorly viable pollen and altered expression of several floral identity genes, among them B-, C- and E-function ones, while no apparent morphological modifications were visible in the other whorls. Oblong ovaries and fruits, that were also parthenocarpic, were obtained in the plants expressing the *TM8*:SRDX repressor gene. Such ovaries showed modified expression of various carpel-related genes. No apparent modifications could be seen in the other flower whorls. The latter plants had also epinastic leaves and malformed flower abscission zones. By using yeast two hybrid assays it was possible to show that *TM8* was able to interact in yeast with MACROCALIX.

Conclusions

The impact of the ectopically altered *TM8* expression on the reproductive structures suggests that this gene plays some role in the development of the tomato flower. MACROCALYX, a putative A-function MADS-box gene, was expressed in all the four whorls of fully developed flowers, and showed quantitative variations that were opposite to those of *TM8* in the anomalous stamens and ovaries. Since the *TM8* protein interacted in vitro only with the Afunction MADS-box protein MACROCALYX, it seems that for the correct differentiation of the tomato reproductive structures possible interactions between *TM8* and MACROCALYX proteins might be important.

Background

The availability of floral homeotic mutants and the isolation of the corresponding genes represented a turning point in our understanding of the molecular basis of flower formation.

These studies were mostly done in Arabidopsis thaliana, Antirrhinum majus and Petunia hybrida, and led to the characterization of various MADS-box transcription factors that were shown to be able to switch on genetic programs leading to the actual formation of the flower organs [1, 2].

Tomato (Solanum lycopersicon) is a species of enormous and worldwide economic importance and a large collection of mutants is presently hosted at the Tomato Genetics Resource Center (http://tgrc.ucdavis.edu). Unfortunately, no tomato floral homeotic mutants were available in 1990′, when the ABC model was proposed, therefore in those years efforts were focused on the isolation of tomato MADS-box genes by screening with heterologous probes two tomato cDNA libraries prepared from mRNA of mature wild type flowers and anantha floral meristems, respectively [3]. In the anantha mutant the floral meristems are blocked before formation of the flower and they branch indefinitely giving rise to a cauliflower-like inflorescence [4].

Several MADS-box coding cDNAs were thus isolated and named TM (Tomato MADS) followed by a number. In particular, from the arrested floral meristem of the anantha mutant it was isolated a gene that was named *TM8*, and was regarded as an "early" gene along the pathway of flower formation together with TM4, while TM5, TM6 and TM16 were regarded as "late" genes along the same pathway [3]. Northern blot assays revealed that *TM8* was expressed in pistils, anthers and petals, although at much lower levels than the other MADSbox genes. No transcripts were detected in sepals and leaves [3] but this result might have depended on the low sensitivity of the Northern technique compared to more modern types of analysis.

In a comprehensive work aimed at the characterization of the MADS-box gene family in tomato, Hileman et al. [5] isolated a large number of new genes, and analyzed by semiquantitative RT-PCR their expression together with that of the previously known genes. It was thus shown that *TM8* is expressed not only in the reproductive apparatus, but also in leaves.

Probably the absence of a *TM8* ortholog in Arabidopsis made this gene of poor interest for further characterization. However, the limited knowledge of the possible function(s) performed by *TM8* was recently stressed by Heijmanset al. [6] who suggested that its functional characterization might be "of special interest in order to complete our understanding of MIKCc gene function".

Matter of fact, after having isolated the first tomato MADS-box genes, the same research group tried to functionally characterize them by preparing transgenic anti-sense plants for each gene. However, to our best knowledge, the only reference to the results obtained with *TM8* is that found in Lifschitz et al. [7] where it was reported that 3 out of 12 transgenic plants exhibited severe deformation of the ovary and complete sterility. Also an extremely high incidence of parthenocarpy was reported.

Since after more than 20 years from the discovery of the *TM8* gene the information about its function is still very scarce, we deemed of interest to study the possible role played by this gene in tomato. To do so we prepared transgenic tomato plants over-expressing it, but also transgenic plants over-expressing a chimeric gene carrying the *TM8* sequence fused to the SRDX transcriptional repressor domain [8,9]. The results obtained in this work indicate that *TM8* may be important for anther but also for ovary and fruit formation. Moreover, some phenotypic alterations were observed also for the leaves and this is in accordance with the observed normal expression of *TM8* in this organ.

Results

A functional characterization of *TM8* was carried out by preparing transgenic plants with altered expression of the gene. Its related cDNA was obtained by RT-PCR experiments using specific oligonucleotides designed on the X60760 sequence [3]. The cDNA thus obtained was sequenced on both strands and it appeared that its coding region was actually 60 nucleotides longer than that of the X60760 sequence. In particular, 42 and 18 extranucleotides were localized in the K and C domains, respectively (not shown). Specific primers able to discriminate between the X60760 sequence and the *TM8* sequence isolated by us were used for PCR experiments with both genomic DNA and cDNA obtained from flowers and fruits. The results (Additional file 1) showed that both sequences are present in the tomato genome, and this is in agreement with data from the published genome [10].

However, only the longer sequence isolated by us appeared to be expressed in the flowers and fruits of our tomato plants (i.e. cv Florida Petite), therefore we decided to use the latter cDNA for the preparation of transgenic Florida Petite plants.

Phenotypic Characterization of Plants Over-expressing the *TM8* Gene

22 independent lines were obtained that harbored the 35S:*TM8* construct as determined by PCR analyses carried out with genomic DNA extracted from leaves (see Methods - generation of transgenic plants). Most of the transgenic lines did not show any macroscopic difference compared to the untransformed ones, however three lines produced flowers with anomalies in the androecia (Figure 1).

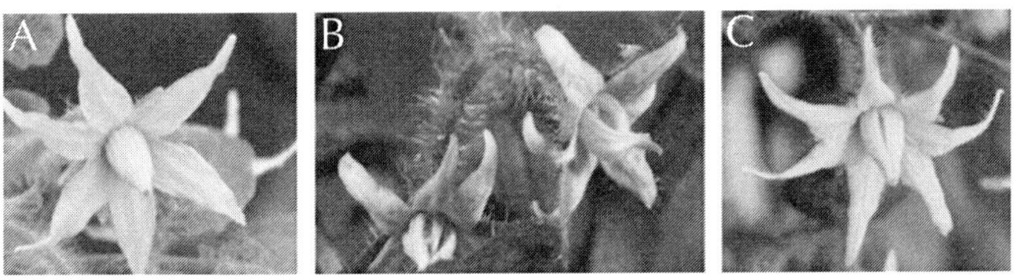

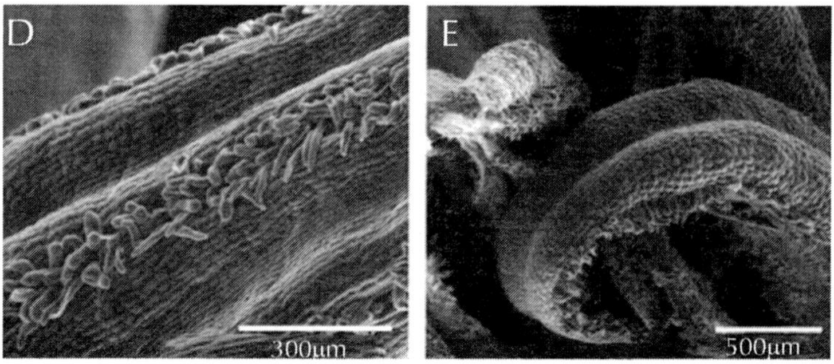

Figure 1: 35S:*TM8* plant phenotype. Wild-type tomato flower (A) and flowers of the lines 35S:*TM8*#16 and 35S:*TM8*#11 (B, C) over-expressing the *TM8* gene and showing splayed out stamens. ESEM (environmental scanning electron microscopy) pictures of a wild-type staminal cone (D) showing the interweaving hairs of the adjoining anthers and a transgenic splayed out cone of the line 35S:*TM8*#16 (E) showing an anther not joined to others.

Normally the tomato anthers form a sort of cone that surrounds the distal part of the style (Figure 1A, 1D). In the anomalous transgenic lines the anthers did not form a regular cone and appeared more (Figure 1B, 1E) or less (Figure 1C) splayed out. Such morphological anomaly suggested that also the anther functionality might have been affected, hence pollen viability assays were performed for the three different lines showing various degrees of splayed out stamens. As it can be seen from Table 1, all three transgenic lines exhibited a highly reduced pollen viability compared to untransformed plants, with line #16 having an extremely low amount (16%) of viable pollen. In accordance with the above data, all the three transgenic lines produced a significantly reduced amount of seeds per fruit, with line #16 yielding only seedless fruits (Table 1). Since the line with the strongest phenotype (i.e. #16) produced seedless fruits and the very little seeds produced by the other two lines were poorly viable, the subsequent molecular characterization could be carried out only with the primary transformants.

Table 1: Pollen viability assay and mean number of seeds per fruit in wild-type and transgenic lines

Plant lines	Pollen viability[a]	Seeds per fruit[b]
wild-type#1	98%	28 ± 6.5
wild-type#2	97%	26.7 ± 6.7
35S:*TM8*#11	25%	5.8 ± 3.4
35S:*TM8*#12	32%	9.3 ± 4.2
35S:*TM8*#16	16%	0 ± 0
35S:*TM8*:SRDX#1	97%	0 ± 0
35S:*TM8*:SRDX#2	96%	0 ± 0
35S:*TM8*:SRDX#6	96%	0 ± 0

[a]Pollen viability assayed according to the MTT test.

[b]Mean number ± standard deviations, Values in boldface are significantly different by Student's t test from wild-type ($P < 0.05$).

Molecular Characterization of Plants Over-expressing the *TM8* Gene

In untransformed flowers the *TM8* gene is generally expressed at very low levels in all four whorls with the highest transcript amount being found in petals, followed by anthers, sepals and ovaries (Figure 2). Analyses were carried out on three different transgenic lines and two untransformed plants (Additional file 2), and all the values shown here represent means of the three transgenic and the two untransformed plants, respectively. The only evident phenotypic effects in the *TM8* over-expressing plants were those found in the stamens (i.e. whorl 3), accordingly the expression profile of MADS-box genes normally expressed in this whorl was studied. Four genes [i.e. TM6, TAP3 (TOMATO APETALA3), SlGLO1 and SlGLO2 (Solanumlycopersicon GLOBOSA 1 and 2)] are known in tomato that belong to the B class [3,11,12].

The over-expression of *TM8* repressed the expression of the four genes in stamens (Figure 3B, C, D, E), while in petals their expression appeared increased (Additional file 3).

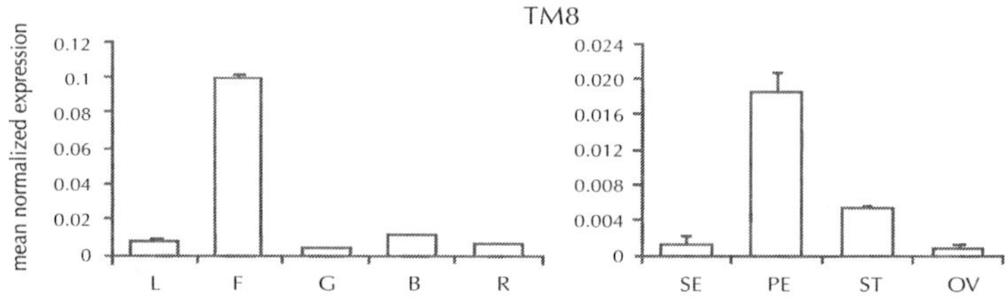

Figure 2: *TM8* gene expression pattern. Relative expression profiles of the *TM8* gene in leaf (L), flower (F), fruits at different developmental stages: green (G), breaker (B), red (R) and flower organs: sepals (SE), petals (PE), stamens (ST) and ovaries (OV). Expression data (means of the normalized expression) were obtained by real-time PCR analyses. Bars are the standard deviations from the means.

Considering the altered expression of the B-function genes in petals, we examined also the expression of MACROCALYX (MC) which is supposed to be an A-function gene in tomato [5,13], and should therefore be expressed in both petals and sepals according to the ABC model of Arabidopsis flower development [1]. As expected, MC was expressed in both sepals and petals, however transcripts were observed also in anthers and ovaries (Figure 3A). Actually, sepals had the highest transcript amount followed by petals, ovaries and anthers, respectively. In the transgenic flowers MC showed unchanged expression levels in sepals and ovaries, while the gene transcripts appeared increased in petals and decreased in the splayed out anthers.

SEPALLATA genes constitute the E function genes and are very important for the formation of all flower whorls since SEP proteins form high order complexes with the ABC transcription factors. In tomato five SEP genes are found, TM5, TM29, SlMBP21 (Solanum lycopersicon MADS-box PROTEIN 21), LeMADS1 (Lycopersicon esculentum MADS1) and RIN (RIPENING-INHIBITOR) [3,13-15], therefore also their expression was analyzed in the *TM8* over-expressing plants. Interestingly, the TM5, TM29 and SlMBP21 genes showed a highly reduced expression in the splayed out anthers compared to the control ones (Figure 3J, K, L). No apparent difference between transgenic and untransformed plants was found in whorls 2, 3 and 4 as regards LeMADS1 (Figure 3M) and RIN, the latter generally expressed at extremely low levels (data not shown).

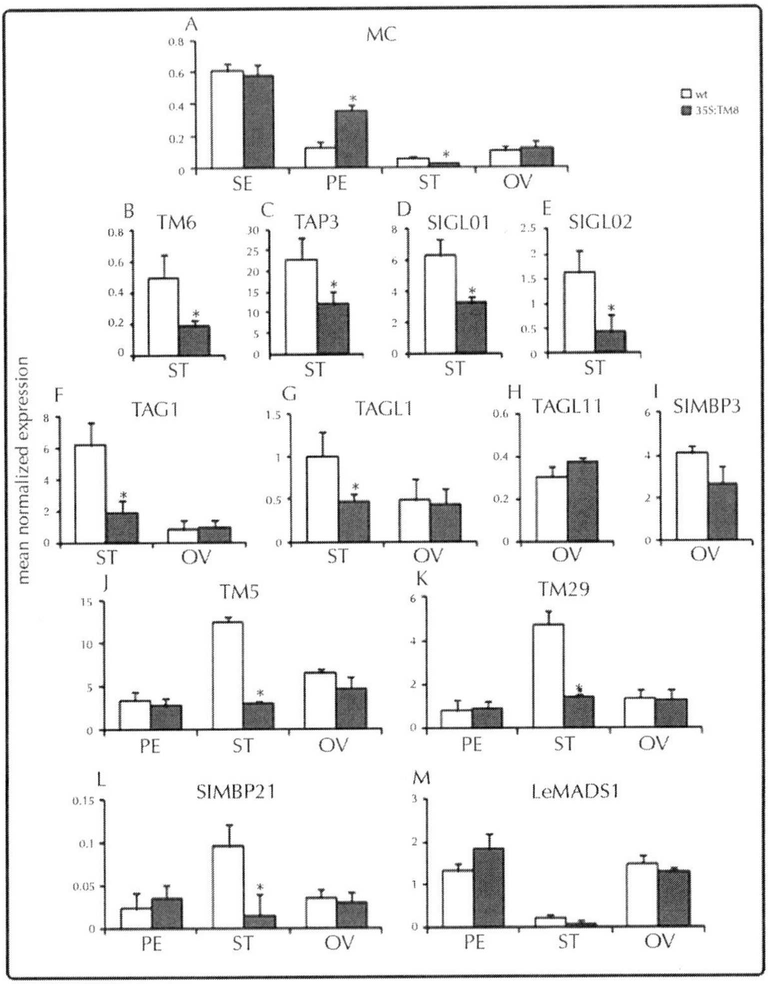

Figure 3: Relative expression of MADS-box genes in 35S:*TM8* plants. Panels from A to M show the relative expression of MADS-box transcription factor encoding genes in wild-type (white) and *TM8* over-expressing (grey) flower organs: (SE), petals (PE), stamens (ST) and ovaries (OV). Each panel incorporates the name of the analyzed gene. Expression data (means of the normalized expression) were obtained by real-time PCR analyses. Values represent the mean of three different transgenic lines and two untransformed plants. Bars are the standard deviations from the means. Asterisks indicate values significantly different by Student's t test from the control (P <0.05).

In tomato four genes [i.e. TAG1 (TOMATO AGAMOUS 1), TAGL1 (TOMATO AGAMOUSLIKE 1), TAGL11 (TOMATO AGAMOUS-LIKE 11) and SLMBP3 (Solanum lycopersicon MADS-box PROTEIN 3)] form the AGAMOUS subfamily [3,15,16] although the genuine C function is represented by the TAG1 and TAGL1 genes while the other two represent the D function genes. Interestingly, no expression difference was found for the C function genes in the transgenic pistils while a significantly decreased expression was found in the transgenic splayed out stamens (Figure 3F, G). As regards the TAGL11 and SLMBP3 genes, they were expressed at high levels in ovaries and no difference was observed between *TM8* overexpressing and untransformed plants (Figure 3H, I).

Phenotypic Characterization of Plants Expressing the *TM8*:SRDX Chimeric Repressor Gene

The expression of the 35S:*TM8*:SRDX construct does actually express the gene in the form of a dominant repressor [8], and this technique has recently been used in tomato to study the role of the TAGL1 MADS-box gene in fruits [17] but also of the Sl-ERF.B3 gene [18]. 15 independent transgenic lines, as judged on the basis of PCR assays carried out on genomic DNA, were obtained and most of them showed visible alterations of the foliage morphology (Figure 4A). In particular, a marked leaf epinasty was evident starting from very young seedlings. Moreover, in all lines the transgenic leaves exhibited a darker green color compared to the untransformed ones, and this was confirmed by measurements of chlorophyll content (Figure 4B).

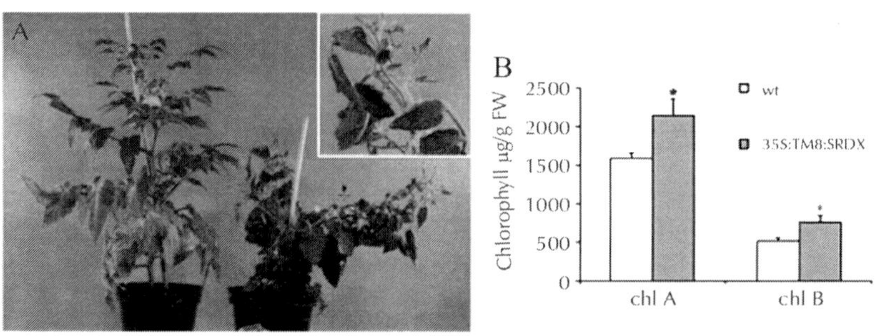

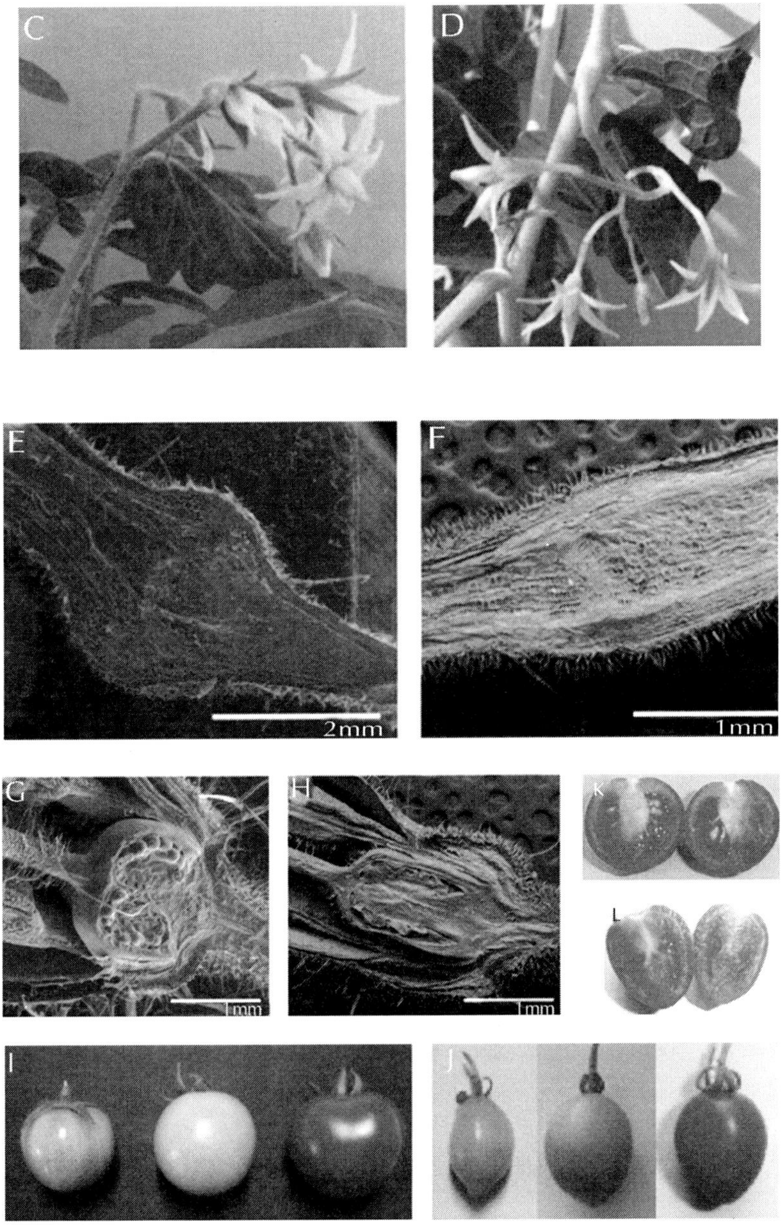

Figure 4: 35S:*TM8*:SRDX plant phenotype. (A) Wild-type tomato plant (left) and plant of the line 35S:*TM8*:SRDX#2 (right) over-expressing *TM8*:SRDX

and having a marked leaf epinasty (see insert). (B) Chlorophyll A and B content in wild-type (white) and 35S:*TM8*:SRDX (grey) leaves. Values represent the mean of three different transgenic lines and two untransformed plants. Bars are the standard deviations from the means. Asterisks indicate values significantly different by Student's t test from the control ($P < 0.05$). Wild-type tomato inflorescence (C) and inflorescence of the line 35S:*TM8*:SRDX#6 (D), the latter having longer flower peduncles. ESEM (environmental scanning electron microscopy) pictures of a wild-type flower abscission zone (E) and of the anomalous flower abscission zone of the line 35S:*TM8*:SRDX#2 (F). ESEM pictures of a wild-type ovary (G) and of the oblong ovary of the line 35S:*TM8*:SRDX#2 (H). Wild-type fruits (I, K) and fruits of the line 35S:*TM8*:SRDX#6 having an oblong morphology (J) and bearing no seeds (L).

Flower and fruit peduncles appeared longer compared to untransformed ones (Figure 4C, D), and the abscission zone did not show a normal organization even when observed by ESEM microscopy (Figure 4E, F). Stamens looked like the wild-type ones, while the fourth whorl of the transgenic plants appeared anomalous. In particular, ovaries and fruits had an oblong form instead of being roundish (Figure 4G, H, I, J), and all the transgenic lines yielded seedless fruits (Figure 4K, L and Table 1) therefore all the subsequent analyses had to be carried out using the primary transformants.

Molecular Characterization of Plants Expressing the *TM8*:SRDX Chimeric Repressor Gene

Analyses were carried out on three different transgenic lines and two untransformed plants, and all the values shown here represent means of the three transgenic and the two untransformed plants respectively. As assessed by real time PCR (Additional file 4) the chimeric repressor transgene was expressed in the transformed plants, therefore the presence of an anomalous phenotype, different from the one induced by the gene over-expression, can be ascribed to the expression of the chimeric repressor.

It is known that in tomato the JOINTLESS (J) MADS-box gene controls the correct formation of the abscission zone in the flower and fruit peduncle [19]. Since in our plants such abscission zone did not show a correct organization, the expression of JOINTLESS was analyzed in these abnormal zones and the transcript amount of this gene resulted significantly reduced compared to the untransformed ones (Figure 5Q).

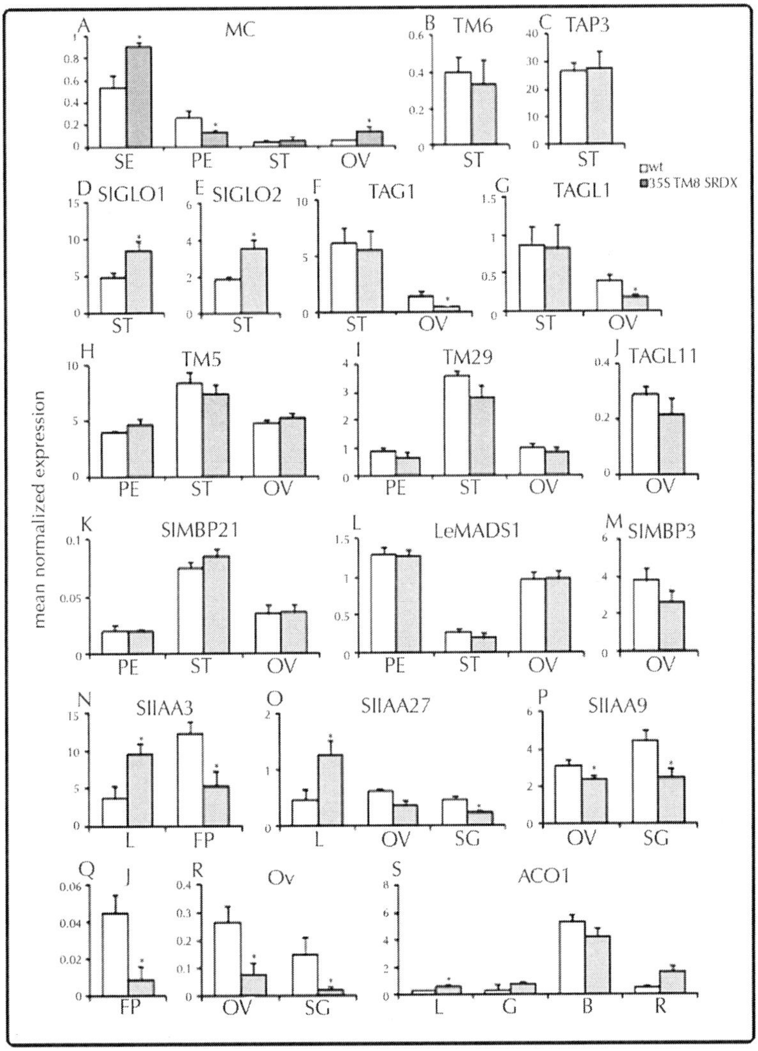

Figure 5: Relative expression of transcription factor encoding genes in 35S:*TM8* plants. Relative expression of different genes in wild-type (white)

and *TM8*:SRDX over-expressing (grey) tissues: sepals (SE), petals (PE), stamens (ST), ovaries (OV), flower peduncles (FP), leaf (L), and fruits at different stages of development [small green (SG), mature green (G), breaker (B) and red (R)]. Panels from A to M show the expression of MADS-box transcription factor encoding genes and each panel incorporates the name of the examined gene. Panels from N to O show the expression of genes coding for different Aux/IAA proteins whose specific name is reported in its dedicated panel. Finally, panel Q, R and S show the expression of the JOINTLESS (J), OVATE (OV) and ACO1 genes, respectively. Expression data (means of the normalized expression) were obtained by real-time PCR analyses. Values represent the mean of three different transgenic lines and two untransformed plants. Bars are the standard deviations from the means. Asterisks indicate values significantly different by Student's t test from the control ($P < 0.05$).

The marked leaf epinasty suggested at first sight an over-production of ethylene by those leaves, yet, in spite of an increased expression of the ACO1 gene (Figure 5S), which is known to be involved in ethylene biosynthesis in tomato [20], no ethylene production could be measured under our experimental conditions. Probably in leaves the production of ethylene was below the sensitivity threshold of the used instrument since the hormone could be measured in the *TM8*:SRDX fruits where it turned out to be produced in amount comparable to that of the untransformed ones (Figure 6).

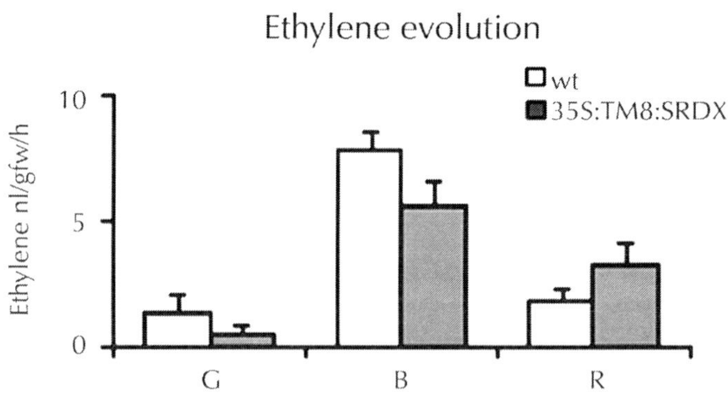

Figure 6: Ethylene evolution. Ethylene evolution in wild-type (white) and *TM8*:SRDX overexpressing (grey) fruits: green (G), breaker (B) and red (R). Values represent the mean of three different transgenic lines and two untransformed plants. Bars are the standard deviations from the means.

In dark-grown seedlings it is known that in the ethylene-induced hook formation a role is played also by auxin, and an interaction between the two hormones has been shown also for leaf epinasty [21]. In order to understand whether auxin might be involved also in the leaf epinasty observed in our transgenic plants it was decided to study the expression of those genes coding for Aux/IAA proteins whose role has already been clarified.

The SlIAA3(Solanum lycopersicon IAA 3) gene has expression regulated by both auxin and ethylene [22]. The down-regulation of this gene reduced the ethylene-induced leaf epinasty, thus suggesting a direct relation between expression of the gene and rate of epinasty [22]. In the transgenic plants with strong epinasty the expression of this gene appeared significantly increased in the leaves while it was reduced in the elongated flower peduncles (Figure 5N). Recently, the gene SlIAA27(Solanum lycopersicon IAA27) coding for another Aux/IAA protein has been characterized [23]. In particular, it has been shown that its silencing leads to the formation of elongated fruits that look like those produced by the 35S:*TM8*:SRDX tomato plants. Moreover, the SlIAA27 silenced plants had leaves with a reduced chlorophyll content. Interestingly, in our transgenic leaves with increased chlorophyll content the SlIAA27 gene shows an increased expression, while in the elongated fruits the transcript amount appears reduced compared to that of the untransformed fruits (Figure 5O).

Regarding SlIAA9 (Solanum lycopersicon IAA9), its down-regulation led to the formation of parthenocarpic fruits [24] similarly to what observed in the*TM8*:SRDX expressing plants. Interestingly, in the latter plants the SlIAA9 gene had a significantly reduced expression in ovaries and very young fruits (10 days after anthesis), that is when seed set is supposed to occur (Figure 5P).

The expression of various MADS-box genes involved in the formation of the reproductive structures was studied in the *TM8*:SRDX expressing plants. Of the four known tomato B class genes, TM6 and TAP3 did not show any significantly modified expression, while both SlGLO1 and SlGLO2 had a significantly increased expression both in stamens (Figure 5B, C, D, E) and in petals (Additional file 5). The expression of the MC gene was unaffected in anthers but not in the other three whorls. The transcript amount appeared increased in sepals and decreased in petals, while in the anomalous ovaries MC had significantly increased expression levels (Figure 5A). Of the four genes belonging to the AGAMOUS subfamily, TAG1 and TAGL1 (the C class genes proper) showed a decreased expression in ovaries and no variation in stamens (Figure 5F, G), while no variations at all was found in these two whorls for TAGL11 and SLMBP3 (Figure 5J, M). Finally, the five SEPALLATA genes did not show any significantly changed expression in whorls 2, 3 and 4 but for a slightly decreased expression of TM29 in the sole transgenic anthers (Figure 5H, I, K, L). The gene OVATE is known in tomato for being involved in the determination of the fruit shape, and its reduced expression could be related to a change in the fruit morphology [25].

The *TM8*:SRDX expressing plants produced elongated ovaries and fruits and the OVATE transcripts were significantly reduced in both ovaries and very small fruits (10 days after anthesis) which are the developmental stages most relevant for OVATE expression and the establishment of the fruit shape (Figure 5R).

Finally, in the elongated fruits both the production of ethylene and the expression of the ACO1 gene were basically comparable to those of the untransformed fruits, thus confirming that the oblong fruits can undergo a normal ripening process (Figure 5S and Figure 6).

MADS-box Protein Interactions

Given the notable involvement of *TM8* in the development of the tomato reproductive structures, we checked the possible ability of the *TM8* protein to interact with other MADSbox transcription factors involved in the development of these structures. To this aim we carried out a large set of tests by means of the yeast two hybrid technique which is a methodology widely used to evaluate the capacity of MADS transcription factors to heteroand /or homo-dimerize in vitro [26,27].

TM8 was not able to form either homo-dimers or complexes with any of the four tomato B type MADS-box transcription factors (TM6, TAP3, SlGLO1 and SlGLO2). The same incapacity to form hetero-dimers was observed when *TM8* was assayed with either JOINTLESS or the three SEPALLATA-like proteins (i.e. TM5, TM29 and SlMBP21) whose cognate genes had shown a markedly affected expression in the flowers with altered *TM8* expression (Table 2). Similarly, *TM8* did not hetero-dimerize with either TAG1 or TAGL1, two tomato C type

MADS-box proteins. Matter of fact, the *TM8*-TAG1 complex could very weakly promote the transcription of just one reporter gene (ADE2), whilst HIS3was not transcribed. Moreover, the weak ADE2 activation was only observed when using *TM8* as bait and TAG1 as prey, therefore these results suggest that this interaction is not solid.

Interestingly, our yeast two-hybrid assays revealed that *TM8* can hetero-dimerize with MC (Table 2). The *TM8*-MC interaction is quite solid since it was detected both with *TM8* as the bait and when MC was the bait protein. Moreover, in both cases the TM8-MC complex was able to activate the transcription of the two reporter genes (HIS3 and ADE2, see Table 2).

Table 2: Protein interactions in yeast

	-W-L[a]	-W-L-H +2 mM 3AT[b]	**-W-L-H +5 mM** 3AT[e]	-W-L-A[d]
TM8-BD/AD	++	--	--	--
TM8-BD/TM8-AD	**++**	--	--	--
BD/TM8-AD	**++**	--	--	--
TM8-BD/MC-AD	++	++	++	++
BD/MC-AD	++			
MC-BD/TM8-AD	++	++	+	+
MC-BD/AD	++	--	--	--
TM8-BD/TM6-AD	++	--	--	--
BD/TM6-AD	++	--	--	--
TM6-BD/TM8-AD	++	--	--	--
TM6-BD/AD	++	--	--	--
TM8-BD/TAP3-AD	++	--	--	--
BD/TAP3-AD	++	--	--	--
TAP3-BD/TM8-AD	++	--	--	--
TAP3-BD/AD	++	--	--	--
TM8-BD/SIGLOI-AD	++	--	--	--
BD/SIGLO I-AD	++	--	--	--
SIGLOI-BD/TMS-AD	++	--	--	--
SIGLOI-BD/AD	++	--	--	--

TM8-BD/SIGL02-AD	++	--	--	--
BD/SIGLO2-AD	++	--	--	--
SIGLO2-BD/TMS-AD	++	--	--	--
SIGLO2-BD/AD	++	--	--	--
TM8-BD/TAG-AD	++	--	--	--
BD/TAG-AD	++	--	--	--
TAG-BD/TM8-AD	++	--	--	--
TAG-BD/TM8-AD	++	--	--	--
TM8-ADITAGL I-BD	++	--	--	--
AD/ TAGL I-BD	++	--	--	--
TM8-BD/1145-AD	++	--	--	--
BD/TM5-AD	++	--	--	--
TM5-BI3/1148-AD	++	--	--	--
TM5-BD/AD	++	--	--	--
TM-8BD/11429-AD	++	--	--	--
BD/TM29-AD	++	--	--	--
TM-29B13/1148-AD	++	--	--	--
TM-29BD/AD	++	--	--	--
TM8-BD/SIMBP2 I-AD	++	--	--	--
BD/SIMBP2 I-AD	++	--	--	--
SIMBP2 I-BD/ TM8-AD	++	--	--	--
SIMBP2 I-BD/AD	++	--	--	--
TM8BD/J-AD	++	--	--	--
8BD/J-AD	++	--	--	--

[a]media lacking either adenine or histidine (supplemented with 3-AT, a competitive inhibitor of the HIS3 enzyme.

[b]-W-L YSD media lacking tryptophan and leucine.

[c]W-L – H YSD media lacking tryptophan, leucine and histidine.

[d]-W-L – A YSD media lacking tryptophan, leucine and adenine.

TM8 has been tested for its ability to form dimers with several tomato MADS-box proteins. Interactions have been assayed using *TM8* both as bait and as prey. All bait and prey chimeric proteins have been controlled, then they have been transformed with the corresponding empty vector in order to exclude auto-activation problems. Yeast transformations have been repeated twice and each time at least four yeast colonies containing a bait and prey plasmids have been tested for their ability to grow in selective media.

Discussion

Transgenic tomato plants with altered expression of the gene were prepared in order to carry out a functional characterization of *TM8*. In particular, in the *TM8* over-expressing plants macroscopic anomalies were found in whorl 3, and they basically consisted of splayed out stamens with poorly viable pollen. Regarding MADS-box genes, splayed out and sterile stamens were first reported by Pnueli et al. [28] in antisense tomato plants with downregulated expression of TM5 which is an E-function gene. In these plants other MADS-box genes, TM6 and TAG1 among them, were unaffected in their expression. Also AmpomahDwamena et al. [14] obtained splayed out and sterile stamens in tomato plants where TM29, another E-function gene, had been ectopically down-regulated, and such down-regulation did not affect the expression of the TM5 and TAG1 MADS-box genes. Finally, splayed out stamens were found in tomato plants with either decreased or missing expression of various B-function genes [12,29].

The above findings are particularly interesting because they show that in the same species both the morphology and the functionality of stamens could be affected in an apparently similar manner by a decreased expression of either one or another of different types of MADS-box genes. Therefore, since the down-regulation of a single gene could cause the same anomalous phenotype, it appears that in tomato the protein products of both the Bfunction genes and the two E-function genes TM5 and TM29 interdependently participate in the process that leads to the differentiation of the third whorl.

The results obtained in this work add further complexity to the molecular network involved in the differentiation of whorl 3 in tomato. In fact, besides the four B-function and the two Efunction genes already mentioned, also SlMBP21, another E-function gene, and both TAG1 and TAGL1, two C-function genes, and MACROCALYX, a putative A-function gene, appeared to be significantly down-regulated in the anomalous splayed out stamens produced by the *TM8* over-expressing tomato plants. Since all the above genes had significantly reduced expression in the anomalous stamens, this finding suggests that in whorl 3 there must be a specific combination and dose equilibrium of various MADS-box proteins in order to have a correct differentiation of stamens. In petals (whorl 2) all the four B-function genes and MACROCALYX appeared significantly up-regulated in the *TM8* over-expressing petals. On the contrary, a mixed situation was found in the *TM8*:SRDX expressing petals: the SlGLO1 and SlGLO2 genes appeared similarly up-regulated and the TM6 and TAP3 genes showed no variations in expression, while MACROCALYX showed a decreased expression. Since no significant morphologic difference could be evidenced in petals of both types of transgenic plants, it appears that the observed changes in the expression of both MACROCALYX and the B-function genes is not sufficient for significantly altering the petal morphology. However, the data regarding the expression of MC in our transgenic plants suggest that this gene must play some role in the differentiation of the whole tomato flower. In fact, even though MC is considered a putative A-function gene, its expression in the fully differentiated flowers was not restricted to the first two whorls, as one would expect on the basis of the canonical ABC model [1]. On the contrary, MACROCALYX was expressed in all 4 whorls, and had a significantly changed expression in both the splayed out anthers and the anomalous ovaries, respectively.

The identity of flower organs is specified by various MIKC MADS-box transcription factors which act in a combinatorial manner [1]. The molecular networks formed by these proteins have been extensively explored using yeast two hybrid assays [26,30]. Such studies have been performed also in tomato [12,15,29], however the possible interactions of the *TM8* protein with other MADS-box transcription factors was never examined. We therefore decided to use this technique to identify those tomato MADS box proteins able to form heterodimers with *TM8*.

Unlike Arabidopsis, tomato has two AP3-like proteins, TAP3 and TM6 [3,11], and two PIlike proteins, SlGLO1 and SlGLO2 [12], which represent the B-class function. None of the B type MADS box proteins was able to form dimers with *TM8* in our yeast two hybrid assays. Yeast two hybrid assays also excluded that *TM8* is able to homo-dimerize. *TM8* did not interact with TAGL1 and TAG1 either. Regarding TAG1, we recorded a weak activation of the ADE2 reporter gene, but we were not able to observe growth on media lacking histidine, which suggests that HIS3 was not activated. Therefore it appears unlikely that *TM8* and TAG1 may form dimers in vivo.

TM8 did not physically interact in yeast with those tomato SEPALLATA-like MADS box protein [TM5, TM29 and SlMP21 [31]] that had altered expression patterns in our transgenic plants, and also with JOINTLESS [19]. Interestingly, *TM8* was able to interact in yeast with MACROCALIX, and the dimer *TM8*-MC could promote the transcription of both reporter genes, ADE2 and HIS3. This confirms that the chimeric *TM8* protein used for yeast assays is properly folded.

Twenty years after the proposal of the ABC model, a modification was introduced by Causier et al. [32] in the review "Floral organ identity: 20 years of ABCs" to account for the absence of an A-function in most plant species. Schwarz-Sommer and co-workers introduced a new (A)-function [33] important to define the floral meristem identity and to produce the sepals that are considered as the ground state of floral organs. Our data seem to suggest that this model might apply also to the tomato flower, and that the activity of the *TM8* protein might be mediated by interactions with the MACROCALYX protein.

The over-expression of the *TM8*:SRDX repressor chimera had macroscopic effects on both reproductive and vegetative structures. Although it has been shown in tomato [5] and in two Gymnosperms [34] that *TM8*-like genes are expressed also in leaves, the latter finding was unexpected because Lifschitz et al. [7] had reported anomalies only for the reproductive structures in their *TM8* antisense tomato plants, while in our transgenic plants the leaves showed a marked epinasty and were greener compared to the untransformed ones. Also the flower peduncles were different compared to wild-types since they did not differentiate a correct abscission zone. To the latter purpose, it is known that a correct expression of the MADS-box gene JOINTLESS is necessary for the differentiation of a normal abscission zone in the tomato flower peduncle [25] and, as expected, in the anomalous abscission zones of the *TM8*:SRDX flower peduncles also the expression of JOINTLESS appeared significantly reduced, in agreement with the defective abscission zones.

In plants the physiological activity of a given hormone may also depend on its interactions with other hormones present in the same tissue, and this has been shown several times for ethylene and auxin [35,36]. In tomato it was demonstrated that SlIAA3, a gene coding for an Aux/IAA protein, can be positively regulated by both auxin and ethylene, and antisense tomato plants for this gene had a reduced epinastic response compared to wild-type ones when treated with exogenous ethylene [36]. Interestingly, in the *TM8*:SRDX expressing plants the epinastic leaves had a significantly increased expression of the SlIAA3 gene, while a significantly reduced transcript amount was found in the lengthened flower peduncles. Therefore, the inability to measure the ethylene produced by the epinastic leaves might simply reflect the need for ethylene to just activate the expression of the SlIAA3gene, therefore the hormone had not to be produced in enormous amounts. Suggestions about a possible auxin involvement came also from other phenotypic characteristics of the 35S:*TM8*:SRDX plants, like the elongated fruits, their parthenocarpy and the deep green color of the foliage. In fact, in these anomalous situations other genes involved in the signal transduction pathway of auxin showed an expression that was altered as expected on the basis of their demonstrated function [22-24]. However, the possible connection between *TM8* and auxin remains elusive.

As regards the phenotypic anomalies of the reproductive structures observed in the *TM8*:SRDX plants, they appeared to affect only whorl 4. In particular, all the fruits were parthenocarpic, a characteristics already described by Lifschitz et al. [7] for their *TM8* antisense tomato plants. On the contrary, the stamens had a normal appearance and the pollen viability was comparable with that of the wild-types, therefore the anomaly was evidently due to problems in the carpel whorl. Actually, transgenic ovaries had an elongated shape that was maintained till the end of their development so that also ripe fruits had an ellipsoidal shape instead of being roundish like the untransformed ones. The shape of tomato fruits is under the control of various genes [37], in particular a low expression of the OVATE gene has been shown to be responsible for the formation of pear-shaped tomatoes [25]. Recently, Rodriguez et al. [37] evidenced that OVATE may also be involved in the formation of ellipsoidal tomatoes, which appears to be the case also for the35S:*TM8*:SRDX fruits since the OVATE gene had a significantly decreased expression in ovaries and very young fruits, that is when the fruit shape is established. The C-function TAG1 gene was shown by Pnueli et al. [38] to be expressed in stamens and carpels, and to be of basic importance for a correct differentiation of these two organs. In particular, they found that a down-regulated expression of the gene caused the appearance of relevant malformations, among which both male and female sterility were reported. In tomato TAG1 and TAGL1 are the genuine C function genes while TAGL11 and SlMBP3 are Dfunction genes. It is interesting to note that the expression of both TAG1 and TAGL1 was consistent with the role played by them during the differentiation of reproductive structures. In the *TM8* over-expressing plants the two genes had significantly decreased expression in the anomalous stamens but not in the normal ovaries, on the contrary in the *TM8*:SRDX expressing plants the two genes had normal expression levels in stamens and significantly reduced expression levels in the anomalous ovaries. Since in the latter ovaries both the Dfunction and the E-function genes did not show any significantly varied expression compared to wild-type, the above data reinforce the role played by the TAG1 and TAGL1 genes in the development of tomato carpels [17,38,39].

CONCLUSIONS

Soon after its discovery, the expression profile of the *TM8* gene was studied by means of a Northern analysis and the Authors found high transcript amounts in the anantha floral meristem where other MADS-box genes (i.e. TM5 and TM6) could not be detected. For this reason the *TM8* gene was defined as an "early" gene and the others as "late" genes along the process of flower differentiation [3]. The above pattern of expression suggests that the early *TM8* gene might, in some yet unknown way, regulate the expression of the late genes, and such an idea appears to be consistent with the results of this work, at least as far as the differentiation of whorls 3 and 4 and the expression of other MADS-box genes are concerned. It will be interesting to study the relations between the expression of *TM8* and the activity of auxin, a hormone that is known to be important for fruit set and development [40]. Recently, Gramzow et al. [41] showed that *TM8*-like genes are quite common also in Gymnosperms. Therefore, in spite of the generally low levels of expression observed for this type of gene, its involvement in the formation of the reproductive structures might be the reason for the widespread conservation of *TM8*-like genes in seed plants.

METHODS

Plant Material

Tomato plants (Lycopersicon esculentum cv. Florida Petite) were grown under standard conditions at 25°C and a 16-h photoperiod in a controlled greenhouse at the Department of Biology, University of Padua. No authorization was needed for growing the tomato plants in the above greenhouse. Seeds were obtained from the Tomato Growers Supply Company, Fort Myers, FL, USA (www.tomatogrowers.com). Flower parts [sepals (SE), petals (PE), stamens (ST), ovaries (OV)] and peduncles (FP) were collected from flowers at anthesis. Fruits were harvested at different developmental stages: small green i.e. 10 days after anthesis (SG), mature green (G), breaker (B) and ripe red (R). Fully expanded leaves (L) were also harvested. All tissues and fruit samples were frozen and stored at −80°C.

RNA Extraction and Gene Expression Analysis

Total RNA was extracted from different tissues according to Chang et al. [42]. RNA yield and purity were checked by means of ultraviolet (UV) absorption spectra, whereas RNA integrity was ascertained by electrophoresis in agarose gel.

The RNA samples obtained from different tissues were converted to cDNA by means of the High-Capacity cDNA Archive Kit (Applied Biosystems,www.lifetechnologies.com), using random hexamers as primers. 3 μg of total RNA, pre-treated with 1.5 U of DNase I (Promega, www.promega.com), were used as starting template. The gene expression analysis was performed by standard real-time PCR. Primer sequences for the selected genes are listed in Additional file 6. The internal standard consisted of the actin gene. PCR was carried out with the Gene Amp 7500 Sequence Detection System (Applied

Biosystems). The obtained CT values were analyzed by means of the Q-gene software by averaging three independently calculated normalized expression values for each sample. Expression values are given as the mean of the normalized expression values of the triplicates, calculated according to equation (2) of the Q-gene software [43].

Generation of Transgenic Plants

The constructs used to produce transgenic plants were prepared using the pBINAr_GWa plasmid obtained by cloning the GWa gateway cassette (Invitrogen, www.lifetechnologies.com) into the pBINAr vector SmaI restriction site, between the 35S CaMV promoter and a nopaline synthase (NOS) terminator [44].

To obtain the 35S:*TM8* construct the *TM8* (TOMATO MADS 8) full length cDNA (accession number KF270624) was PCR-amplified using primers (FW 5'- CATTTGAAGAATGGGGAGAG - 3' and RV 5'-AGGTAGCAATTGAAGCTCTG - 3') designed on the already available *TM8* sequence (X60760) and was subsequently cloned into the pCR®8/GW/TOPO® vector (Invitrogen).

Using the Gateway LR Clonase enzyme mix (Invitrogen) the *TM8* cDNA was cloned into the pBINAr_GWa plasmid via homologous recombination.

A dominant repressor construct (35S:*TM8*:SRDX) was created by generating a translational fusion between the EAR repression domain (SRDX) [8] and the 3' end of the *TM8* cDNA by means of an RT-PCR experiment using the primers FW 5'- CATTTGAAGAATGGGGAGAG - 3' and RV 5'- TTTTAAGC-GAAACCCAAACGGAGTTCTAGATCCAGATCGAGTCCCTTAGAAAGTA ACTC-3' (the latter containing the SRDX repression domain). Subsequently the *TM8*:SRDX amplicon was introduced into the pBINAr_GWa as described above.

The identity of the cloned cDNAs was ascertained by sequencing. DNA sequencing was performed by BMR Genomics, Padua, Italy (www.bmr-genomics.it). Sequence manipulations, analyses, and alignments were performed using the LASERGENE software package (DNASTAR, www.dnastar.com).

The resulting binary plasmids were inserted in Agrobacterium tumefaciens (strain LBA4404) cells that were then used to transform tomato according to Fillati et al. [45]. Kanamycinresistant plants were confirmed for the presence of the transgene by means of PCR, using a primer on the cauliflower mosaic virus (CaMV) 35S promoter (5'- GGGGAATTCGGTGGCTCCTACAAATG- 3') and a primer on the *TM8* coding sequence (for the 35S:*TM8* construct: 5'- TCATCCCTTAGAAAGTAACTCACT- 3' and for the 35S:*TM8*:SRDX : 5' - TTCTAGATCCAGATCGAGTCCCTTAG- 3').

Pollen Viability Test

Pollen viability was ascertained using the MTT [MTT: 3-(4,5-dimethylthiazolyl-2)-2,5- diphenyltetrazoliumbromide] assay according to [46]. Briefly, the anthers of ten flowers taken from the plant of interest were introduced into a solution containing MTT 1% (w/v) (Sigma-Aldrich, www.sigmaaldrich.com) and sucrose 5% (w/v). After ten minutes 1000 pollen grains per plant were counted under an optical microscope (LEICA DM5000, www.leica-microsystems.com/). Pollen grains were considered viable if they turned deep pink. Wild-type anthers incubated for 2 hours at 80°C were used as negative control.

Microscopy Analysis

Tomato tissues (flower parts and peduncles) were observed without any treatment under lowpressure conditions by means of environmental scanning electron microscopy (ESEM) at the CUGAS facilities, University of Padua (www.unipd.it/cugas/).

Ethylene Measurement

Ethylene production was determined by enclosing either the whole fruits or the leaves in jars (50–60 ml, according to need), sealed with a transparent wrapper and kept in the light at room temperature. After 1 h, 1 ml of air sample was withdrawn from each jar for the ethylene measurements. A gas chromatograph (Perkin–Elmer F17; Norwalk, Conn, www.perkinelmer.com) was used for quantifying ethylene concentrations.

Yeast Two-hybrid and Ternary-trap Assays

YPAD, SD, and appropriate dropout media have been described previously [47]. The yeast two-hybrid assays were performed using the yeast strain AH109 (Clontech, www.clontech.com) [48]. The pGBTKT7 vector (Clontech), carrying the GAL4 DNAbinding domain, was used to express the bait proteins, whereas the pGADT7 vector [49], carrying the GAL4 activation domain, was used to express the prey proteins. Two-hybrid protein interactions were evaluated by growing the yeast colonies at 28°C on media lacking either histidine or adenine and supplemented with different amounts of 3-amino-1,2,4-triazole (3-AT).

All the primers used to clone the tomato MADS box genes are listed in the Additional file 7. Accession number The *TM8* sequence used in this work will appear in Genbank under the following accession number: KF270624.

Competing Interests

The authors declare that they have no competing interests.

Authors' Contributions

MD prepared the constructs and made the transgenic plants, carried out their phenotypic characterization, and did molecular analyses; SM and FR prepared constructs and carried out the two-hybrid assays and interpreted the results; AL carried out molecular analyses; GC conceived the research, coordinated it and wrote the paper. All the Authors critically read and approved the final version of the manuscript.

ACKNOWLEDGMENTS

G.C. was supported by a grant from Ministero dell'Istruzione, dell'Università e della Ricerca (MIUR), Italy. S.M. and F.R. were supported by the Cariplo Foundation (SeeFruit 2011–2257).

We thank Dr A. Botton for assistance with ethylene measurements and Dr A. Pavanello for technical help.

REFERENCES

1. Coen ES, Meyerowitz EM: The war of the whorls: genetic interactions controlling flower development. Nature 1991, 353:31–37.
2. Becker A, Theissen G: The major clades of MADS-box genes and their role in the development and evolution of flowering plants. Mol Phylogenet Evol 2003, 29:464–489.
3. Pnueli L, Abu- Abeid M, Zamir D, Nacken W, Schwarz-Sommer Z, Lifschitz E: The MADS box gene family in tomato: temporal expression during floral development, conserved secondary structures and homology with homeotic genes from Anthirrinum and Arabidopsis. Plant J 1991, 1:255–266.
4. Helm J: Vergleichende Betrachtungenuber die Entwicklung der Enfloreszenz bei Lycopersicum esculentum Mill. Und bei einerRontgen mutante. Zeuchter 1951, 21:89– 95.
5. Hileman LC, Sundstrom JF, Litt A, Chen M, Shumba T, Irish VF: Molecular and phylogenetic analyses of the MADS-box gene family in tomato. Mol Biol Evol 2006, 23:2245–2258.
6. Heijmans K, Morel P, Vandenbussche M: MADS-box genes and floral development: the dark side. J Exp Bot 2012, 63:5397–5404.
7. Lifschitz E, Brodai L, Hareven D, Hurwitz C, Prihadash A, Pnueli L, Samach A, Zamir D: Molecular mapping of flower development in tomato. In Molecular Biology of Tomato. Edited by Yoder J. Lancaster PA, USA: Technomic Publishing Company Incorporated; 1993:175–184.
8. Hiratsu K, Matsui K, Koyama T, Ohme-Takagi M: Dominant repression of target genes by chimeric repressors that include the EAR motif, a repression domain, in Arabidopsis. Plant J 2003, 34:733–739.
9. Matsui K, Ohme-Takagi M: Detection of protein-protein interactions in plants using the transrepressive activity of the EAR motif repression domain. Plant J 2010, 61:570– 578.
10. Tomato Genome Consortium: The tomato genome sequence provides insights into fleshy fruit evolution. Nature 2012, 485:635–641.
11. Kramer EM, Dorit RL, Irish VF: Molecular evolution of genes

controlling petal and stamen development: duplication and divergence within theAPETALA3 and PISTILLATA MADS-Box gene lineages. Genetics 1998, 149:765–783.

12. Geuten K, Irish V: Hidden variability of floral homeotic B genes in Solanaceae provides a molecular basis for the evolution of novel functions. Plant Cell 2010, 22:2562– 2578.
13. Vrebalov J, Ruezinsky D, Padmanabhan V, White R, Mediano D, Drake R, Schuch W, Giovannoni JJ: A MADS-box gene necessary for fruit ripening at the tomato ripeninginhibitor (rin) locus. Science 2002, 296:343–346.
14. Ampomah-Dwamena C, Morris BA, Sutherland P, Veit B, Yao JL: Down-regulation of TM29, a tomato SEPALLATA homolog, causes parthenocarpic fruit development and floral reversion. Plant Physiol 2002, 130:605–617.
15. Leseberg CH, Eissler CL, Wang X, Johns MA, Duvall MR, Mao L: Interaction study of MADS-domain proteins in tomato. J Exp Bot 2008, 59:2253–2265.
16. Busi MV, Bustamante C, D'Angelo C, Hidalgo-Cuevas M, Boggio SB, Valle EM, Zabaleta E: MADS-box genes expressed during tomato seed and fruit development. Plant Mol Biol 2003, 52:801–815.
17. Itkin M, Seybold H, Breitel D, Rogachev I, Meir S, Aharoni A: TOMATO AGAMOUSLIKE 1 is a component of the fruit ripening regulatory network. Plant J 2009, 60:1081– 1095.
18. Liu M, Pirrello J, Kesari R, Mila I, Roustan JP, Li Z, Latché A, Pech JC, Bouzayen M, Regad F: A dominant repressor version of the tomato Sl-ERFB3 gene confers ethylene hypersensitivity via feedback regulation of ethylene signaling and response components. Plant J 2013, 76:406–419.
19. Mao L, Begum D, Chuang HW, Budiman MA, Szymkowiak EJ, Irish EE, Wing RA: JOINTLESS is a MADS-box gene controlling tomato flower abscission zone development. Nature 2000, 406:910–913.
20. Barry CS, Blume B, Bouzayen M, Cooper W, Hamilton AJ, Grierson D: Differential expression of the 1-aminocyclopropane-1-carboxylate oxidase gene family of tomato. Plant J 1996, 9:525–535.

21. Lin Z, Arciga-Reyes L, Zhong S, Alexander L, Hackett R, Wilson I, Grierson D: SlTPR1, a tomato tetratricopeptide repeat protein, interacts with the ethylene receptors NR and LeETR1, modulating ethylene and auxin responses and development. J Exp Bot 2008, 59:4271–4287.
22. Chaabouni S, Jones B, Delalande C, Wang H, Li Z, Mila I, Frasse P, Latché A, Pech JC, Bouzayen M: Sl-IAA3, a tomato Aux/IAA at the crossroads of auxin and ethylene signalling involved in differential growth. J Exp Bot 2003, 60:1349–1362.
23. Bassa C, Mila I, Bouzayen M, Audran-Delalande C: Phenotypes associated with downregulation of Sl-IAA27 support functional diversity among Aux/IAA family members in tomato. Plant Cell Physiol 2012, 53:1583–1595.
24. Wang H, Jones B, Li Z, Frasse P, Delalande C, Regad F, Chaabouni S, Latché A, Pech JC, Bouzayen M: The tomato Aux/IAA transcription factor IAA9 is involved in fruit development and leaf morphogenesis. Plant Cell 2005, 17:2676–2692.
25. Liu J, Van Eck J, Cong B, Tanksley SD: A new class of regulatory genes underlying the cause of pear-shaped tomato fruit. Proc Natl Acad Sci U S A 2002, 99:13302–13306.
26. Davies B, Egea-Cortines M, de Andrade SE, Saedler H, Sommer H: Multiple interactions amongst floral homeotic MADS box proteins. EMBO J 1996, 15:4330–4343.
27. Masiero S, Li MA, Will I, Hartmann U, Saedler H, Huijser P, Schwarz-Sommer Z, Sommer H: INCOMPOSITA: a MADS-box gene controlling prophyll development and floral meristem identity in Antirrhinum. Development 2004, 131:5981–5990.
28. Pnueli L, Hareven D, Broday L, Hurwitz C, Lifschitz E: The TM5 MADS box gene mediates organ differentiation in the three inner whorls of tomato flowers. Plant Cell 1994, 6:175–186.
29. de Martino G, Pan I, Emmanuel E, Levy A, Irish VF: Functional analyses of two tomato APETALA3 genes demonstrate diversification in their roles in regulating floral development. Plant Cell 2006, 18:1833–1845.
30. de Folter S, Immink RG, Kieffer M, Parenicová L, Henz SR, Weigel D, Busscher M, Kooiker M, Colombo L, Kater MM, Davies B, Angenent GC: Comprehensive interaction map of the Arabidopsis

MADS Box transcription factors. Plant Cell 2005, 17:1424–1433.

31. Liu D, Wang D, Qin Z, Zhang D, Yin L, Wu L, Colasanti J, Li A, Mao L: The SEPALLATA MADS-box protein SLMBP21 forms protein complexes with JOINTLESS and MACROCALYX as a transcription activator for development of the tomato flower abscission zone. Plant J 2014, 77:284–296.
32. Causier B, Schwarz-Sommer Z, Davies B: Floral organ identity: 20 years of ABCs. Semin Cell Dev Biol 2010, 21:73–79.
33. Schwarz-Sommer Z, Huijser P, Nacken W, Saedler H, Sommer H: Genetic control of flower development by homeotic genes in Antirrhinum majus. Science 1990, 250:931– 936.
34. Lovisetto A, Guzzo F, Tadiello A, Toffali K, Favretto A, Casadoro G: Molecular analyses of MADS-box genes trace back to Gymnosperms the invention of fleshy fruits. Mol Biol Evol 2012, 29:409–419.
35. Lehman A, Black R, Ecker JR: HOOKLESS1, an ethylene response gene, is required for differential cell elongation in the Arabidopsis hypocotyl. Cell 1996, 85:183–194.
36. Chaabouni S, Latché A, Pech JC, Bouzayen M: Tomato Aux/IAA3 and HOOKLESS are important actors of the interplay between auxin and ethylene during apical hook formation. Plant Signal Behav 2009, 4:559–560.
37. Rodríguez GR, Muños S, Anderson C, Sim SC, Michel A, Causse M, McSpadden Gardener BB, Francis D, van der Knaap E: Distribution of SUN, OVATE, LC, and FAS in the tomato germplasm and the relationship to fruit shape diversity. Plant Physiol 2011, 156:275–285.
38. Pnueli L, Hareven D, Rounsley SD, Yanofsky MF, Lifschitz E: Isolation of the tomato AGAMOUS gene TAG1 and analysis of its homeotic role in transgenic plants. Plant Cell 1994, 6:163–173.
39. Vrebalov J, Pan I, Arroyo AJM, McQuinn R, Chung M, Poole M, Rose J, Seymour G, Grandillo S, Giovannoni J, Irish VF: Fleshy fruit expansion and ripening are regulated by the tomato SHATTERPROOF gene TAGL1. Plant Cell 2009, 21:3041–3062.
40. Spena A, Rotino GL: Parthenocarpy. State of the Art. In Current Trends in the Embryology of Angiosperm. Edited by Bhojwani

SS, Soh WY. The Netherlands: Kluwers Academic Publishers; 2001:435–450.

41. Gramzow L, Weilandt L, Theißen G: MADS goes genomic in conifers: towards determining the ancestral set of MADS-box genes in seed plants. Ann Bot-London 2014, 114:1407–1429.
42. Chang S, Puryear J, Cairney J: A simple and efficient method for isolating RNA from pine trees. Plant Mol Biol Rep 1993, 11:113–116.
43. Muller PY, Janovjak H, Miserez AR, Dobbie Z: Processing of gene expression data generated by quantitative realtime RT-PCR. Biotechniques 2002, 32:1372–1379.
44. Liu XJ, Prat S, Willmitzer L, Frommer WB: Cis regulatory elements directing tuberspecific and sucrose-inducible expression of a chimeric class I patatin promoter/GUSgene fusion. Mol Gen Genet 1990, 223:401–406.
45. Fillati JJ, Kiser J, Rose R, Comai L: Efficient transfer of a glyphosate tolerance gene into tomato using a binary Agrobacterium tumefaciens vector. Nat Biotechnol 1987, 5:726–730.
46. Wang ZY, Ge Y, Scott M, Spangenberg G: Viability and longevity of pollen from transgenic and non transgenic tall fescue (Festuca arundinacea) (Poaceae) plants. Am J Bot 2004, 91:523–530.
47. Sherman F: Getting started with yeast. Methods Enzymol 1991, 194:3–21.
48. James P, Halladay J, Craig EA: Genomic libraries and a host strain designed for highly efficient two-hybrid selection in yeast. Genetics 1996, 144:1425–1436.
49. Harper JW, Adami GR, Wie N, Keyomarsi K, Elledge SJ: The p21 Cdk-interacting protein Cip1 is a potent inhibitor of G1 cyclin-dependent kinases. Cell 1993, 75:805–816.

Citations

CHAPTER 1

FABIO R. SCARANO, Structure, Function and Floristic Relationships of Plant Communities in Stressful Habitats Marginal to the Brazilian Atlantic Rainforest, Ann Bot (2002) 90 (4): 517-524 first published online September 4, 2002 doi:10.1093/aob/mcf189.

CHAPTER 2

MARIA ROSÂNGELA SIGRIST and MARLIES SAZIMA, Pollination and Reproductive Biology of Twelve Species of Neotropical Malpighiaceae: Stigma Morphology and its Implications for the Breeding System, Ann Bot (2004) 94 (1): 33-41 first published online June 11, 2004 doi:10.1093/aob/mch108

CHAPTER 3

Walsh RP, Arnold PM, Michaels HJ. 2014. Effects of pollination limitation and seed predation on female reproductive success of a deceptive orchid. AoB PLANTS 6: plu031; doi:10.1093/aobpla/plu031.

CHAPTER 4

Caperta AD, Espı´rito-Santo MD, Silva V, Ferreira A, Paes AP, Ro´is AS, Costa JC, Arse´nio P. 2014. Habitat specificity of a threatened and endemic, cliff-dwelling halophyte. AoB PLANTS 6: plu032; doi:10.1093/aobpla/plu032.

CHAPTER 5

Armbruster WS. 2014. Floral specialization and angiosperm diversity: phenotypic divergence, fitness trade-offs and realized pollination accuracy. AoB PLANTS 6: plu003; doi:10.1093/aobpla/plu003.

CHAPTER 6

Juliana Lischka Sampaio Mayer, Sandra Maria Carmello-Guerreiro, and Paulo Mazzafera, A functional role for the colleters of coffee flowers, doi: 10.1093/aobpla/plt029.

CHAPTER 7

Yash Mangla, Rajesh Tandon, Shailendra Goel and S. N. Raina, Structural organization of the gynoecium and pollen tube path in Himalayan sea buckthorn, Hippophae rhamnoides (Elaeagnaceae) doi: 10.1093/aobpla/plt015

CHAPTER 8

Bramwel W. Wanjala, Meshack Obonyo, Francis N. Wachira, Alice Muchugi, Margaret Mulaa, Jagger Harvey, Robert A. Skilton, Janice Proud and Jean Hanson, genetic diversity in napier grass (pennisetum purpureum) cultivars: implications for breeding and conservation, doi: 10.1093/aobpla/plt022.

CHAPTER 9

Michelle L. Jones, mineral nutrient remobilization during corolla senescence in ethylene sensitive and -insensitive flowers, doi: 10.1093/aobpla/plt023

CHAPTER 10

Margherita Daminato, Simona Masiero, Francesca Resentini, Alessandro Lovisetto, and Giorgio Casadoro, Characterization of TM8, a MADS-box gene expressed in tomato flowers, BMC Plant Biology 2014, 14:319 doi:10.1186/s12870-014-0319-y.

Index

A

Aflp Analysis 221
Aflp Marker, Reliability 230
Aflp Primer Combination 222
Agamospermous Species 37
Agamospermy 34, 37
Amplified Fragment Length Polymorphism (AFLP) 216, 217
Analysis of Molecular Variance (AMOVA) 222
Ancestors 114
Androecia 27
Angiosperm 194, 195, 206, 207, 209, 210, 212, 213
Angiosperm Flower 238
Angiosperms 51, 109, 128, 129, 139, 154, 156, 160, 163, 165, 166
Animals 109, 111, 115, 126, 129, 132
Arabidopsis 249, 250, 251, 252, 256, 257, 260, 261

B

Biodiversity 1, 2, 15, 22
Biogeographic Position 81
Bitegmic Ovule 200
Botanical Material 174
Breaker 270, 276, 286
Breeding Systems 26, 27, 29, 45, 81
Broad Promontory 83

Bud Flower, Morphology 176

C

Capsules 56, 57, 58, 59, 60, 61, 66
C. Arabica 172, 184, 186, 187
Caryocar Brasiliense Camb 171
Catuaí 170, 173, 174, 176, 177, 179, 180, 181, 182, 183, 184, 185, 186, 187
Catuaí Colleter 170, 182
Cetyltrimethylammonium Bromide (CTAB) 221
Chimeric Gene 266
Cluster 80, 87, 89, 93, 94
Clusters 90
Coastal Plants 94
Coffea Arabica 170, 171, 174, 188, 189, 190, 191
Conduplicated Margin 194
Cross-Pollination 172

D

Decaffeinated Mutant 172
Decaffito 170, 172, 173, 174, 176, 177, 178, 179, 180, 183, 184, 185, 186, 187, 190
Decaffito Flower 176, 185, 186
Decaffito Mutant 170, 174
Diameter 32
Dictyosome 181, 182, 186
Dna Extraction 221
Dna Fragmentation 239, 258
Dry Weight (Dw) 240

E

Ecological Plasticity 9, 16
Ecosystems 27
Elaiophores 27, 33, 41
Epifluorescence Photomicrograph 200
Epifluorescence Photomicroscope 197
Erysimum Linifolium 250
Ethylene Measurement 288
Extracellular Matrix (ECM) 203

F

Flag Petal 34, 36
Floristic Groups 89
Flowering Plants 110, 126, 127, 133, 141, 155
Flowering Stems 50, 55, 57, 58, 59, 60, 61, 65, 66, 68
Fresh Weight (Fw) 240
Fruit Abortion 56, 57
Fruit Production 60, 64, 65, 69, 76
Fungi 54

G

Gene Expression Pattern 238, 270
Generalized Linear Model (GLM) 57
Genetic Data 96
Genetic Diversity, Evaluation 223, 231
Genetic Diversity Indices 219
Genome 267, 290
Geophyte Palm 12, 15
Germplasm 216, 217, 218, 230, 231, 233, 234
Gladiolus 247
Global Positioning System (GPS) 87
Glycol Methacrylate Monomer

Mixture 196
Gold–Palladium 28
Gradualistic Principle' 133
Grasslands 5
Gynodioecism 96
Gynoecia 27
Gynoecium, Structural Organization 198, 206

H

Hermaphroditism 34
Hibiscus Rosa-Sinensis 246, 260
Hippophae Rhamnoide 194, 195, 296
Homomorphic Approach Herkogamy' 34
Hummingbirds 116

I

Integrity 130, 132
International Livestock Research Institute Forage Germplasm (ILRI-FG) 218
Isolation 109, 111, 113, 114, 123, 125, 128, 129, 130, 131, 132, 137, 138, 142, 143, 144, 145, 150, 152, 153, 156, 157, 160, 162, 164, 166

L

Landscape 65
Large Bees 116, 118, 119, 120, 137
Legal Protection 97
Less Energy 67
Light Microscopy 174
Logarithmic Transformation 88
Low Frequency 94
Lpomoea Purpurea 246, 259, 261

M

Macroevolutionary Scale 146
Magnifying Glass 28, 29
Mating System 66
Microevolutionary 110, 111, 150
Mitchell Diploid (MD) 242
Morphology 26, 27, 28, 32, 43, 45, 115, 116, 121, 134, 135, 136, 139, 145, 147, 148, 154, 160, 161, 162
Mountain Peaks 10
Mutagenize Coffee Seed 172

N

Narcissus Pseudonarcissus 250, 257
Narrow-Tube 116
National Agricultural Research Organization (NARO) 223
Nei's Genetic Distance 224
Nopaline Synthase (NOS) 287
Nurse Plants 2, 6, 11, 12, 13, 14, 15, 16, 18
Nutrient Remobilization 246, 248, 249

O

Offspring 52, 54, 66, 68
Oligarchy 10
Ombrotypes 84
Orchard Grass 232
Original Area 2
Ovary 264, 265, 269, 270, 271, 272, 274, 276, 277, 278, 282, 284, 285, 286

P

Petal 265, 269, 270, 271, 276, 277, 282, 286
Phenetic Analysis 223
Physiognomy 10, 12
Physiology 9
Phytogeographic Study 9
Plant Communities 1, 2, 3, 4, 5, 13, 15, 16
Plant Diversity 81
Plant Reproduction 50
Plant Size 51, 54, 56, 59, 60, 61, 64, 67
Plant Speciation 9
Plant Species 5, 8, 9, 12, 14, 20, 81, 95, 111, 129, 131, 132, 140, 144, 162
Pollen Source 50, 61
Pollen Tube Path 194, 195, 197, 201, 202, 204, 206, 207, 208, 296
Pollen Viability Assay 269
Pollen Viability Test 288
Pollination Strategy 50, 51, 68
Pollination System 27
Pollinators 50, 51, 52, 58, 64, 65, 66, 67, 70, 71, 75, 108, 110, 111, 112, 113, 115, 116, 119, 120, 122, 123, 124, 125, 126, 127, 128, 129, 130, 131, 132, 133, 134, 135, 136, 137, 138, 139, 140, 141, 142, 143, 144, 145, 149, 151, 152, 153, 155, 157, 165
Polymerase Chain Reaction (PCR) 222
Polymorphisms 96
Polyvinylpyrrolidone (PVP) 221
Population 51, 52, 53, 54, 55, 56, 64, 66, 69, 72, 80, 81, 82, 83, 88, 94, 96, 97, 98, 105
Population Genetic Structure 231
Progamic Phase 197, 202, 207

Q

Quaternary 5, 7, 16, 17, 20

R

Rainfall 5, 6
Random Amplified Polymorphic Dna (RAPD) 217
Reinforcing 50, 66, 67
Restriction Fragment Length Polymorphism (RFLP) 217

S

Scanning Electron Micrograph 177, 178
Scanning Electron Microscopy 175
Scree Slopes 94
Scrutiny 109
Secretory Cell 170, 175, 177, 180, 181, 182, 183, 184, 186, 187
Secretory Stage 182, 183
Seed Predators 50, 52, 54, 66, 67, 70, 71
Seed Screening 96
Sepal 265, 269, 270, 276, 277, 283, 286
Simple Sequence Repeat (SSR) 217
Sodium Cacodylate Buffer 196
Species 80, 81, 82, 84, 87, 88,

89, 90, 91, 93, 94, 95, 97, 98, 99, 100, 101, 102, 103, 104, 105, 106
Stamen 264, 265, 268, 269, 270, 271, 272, 276, 277, 278, 281, 282, 284, 285, 286
Stems And Leaves 51, 67
Stigma–Pseudostyle 199, 200, 201, 205, 208
Stipule 171, 184, 190
Subdermal Tissue 202
Sweet Odour 32
Syndromes 109, 133, 149, 154, 159

T

Temperatures 2, 3, 6, 10, 12
Thin Soil 82
Tools For Population Genetic Analysis (TFPGA) 222
Transgenic Florida Petite Plant 267
Transgenic Line 267, 268, 269, 271, 272, 274, 276
Transgenic Petunias 238, 248, 258
Transporter Protein (TP) 240
Tree Species 9, 11

U

Ultraviolet (UV) 286
Universidade Estadual De Campinas (UEC) 28
Unpollinated Gynoecia 196
Unpollinated Stigma 199
Unweighted Pair Group Method With Arithmetic Mean (UPGMA) 222
Upgma Dendrogram 227
Upgma Neighbour-Joining Dendrogram 228
Urban Area 28

V

Vegetative Stem 53

W

Woody Species 12

Y

Yellow Stripe-Like (YSL) 249